Faces Of The Distinguished Flying Cross Of Central Florida

In Their Own Words

2nd Edition

Joseph R. Finch et al.

Faces of the Distinguished Flying Cross of Central Florida

In Their Own Words

2nd Edition

COPYRIGHT INFO

ISBN: 978-1735603704 (hardcover)

ISBN: 978-1735603711 (softcover)

Printed 2020, 2021

This 2nd Edition contains tributes to those who passed away since publication of the first book in 2017. It contains nine new chapters of men who recently joined the Distinguished Flying Cross Society, The Villages Chapter, including the chapter from Don Procter who earned a Distinguished Flying Cross in the Global War on Terror.

Authored by LTC Joseph R. Finch; MAJ Eugene (Barney) Barnhart; CW2 Richard Burns; LTC Jack Deal; LTC Dennis P. Derrig; MAJ Michael Gulick (President, The Villages Chapter, DFC Society); CPT James A. Howd; LTC Fred Koerner; LTC David E. Lehtonen; LT Wayne Paul Lotsberg; LTC Dwight Moore; CPT John F. Morgan; SSGT Gregg Nemeth; CW4 Donald Procter; LTC Ron Schwenkler; COL Dennis P. Sharon; CPT Robert Sherman; CW2 Gary Spooner; LTC Dean R. Vetter; CPT John Webb; and CPT Richard (Dick) Yocca.

Most images provided by individual contributors as part of their own story. Otherwise images are from the public domain or attribution is provided in place. Selected artistic renderings and graphics provided by Don Finch.

Unless otherwise noted Scripture quotations are from The Holy Bible, King James Version.

Table of Contents

FRONTISPIECE

FOREWORD

"...for heroism or extraordinary achievement while participating in an aerial flight."

The Distinguished Flying Cross (DFC) medal is awarded to aviators and aircrew for heroism or extraordinary achievement during aerial flight. The DFC is conferred by all five military services, in all wars and campaigns, from World War I to the present.

The DFC arose out of the First World War, when military aviation was first introduced to combat and the complexities of this task became apparent. The award was established by Congress in 1926 to be the highest military decoration awarded to any officer or enlisted member of the United States Armed Forces who distinguished himself or herself in support of operations by "heroism or extraordinary achievement while participating in an aerial flight."

The DFC was designed by Elizabeth Will and Arthur E. DuBois. It includes a bronze cross pattée with four propeller blades on the obverse and rays between the arms of the cross.

President Coolidge presented the first DFC medal on 11 June, 1927 to Captain Charles A. Lindbergh of the Army Corps Reserve for his solo flight of 3600 miles across the Atlantic Ocean in the Spirit of St. Louis. Other notable recipients of the DFC include Wilbur and Orville Wright, who received the DFC in 1928 for achieving the first manned powered flight. In 1932, Amelia Earhart became the first woman, and the first civilian, to receive the DFC for her solo flight across the Atlantic Ocean. Former President George H. W. Bush was honored with the DFC for piloting a Grumman TBM Avenger on September 2, 1944 during World War II.

This book seeks to honor 33 recipients of the DFC from the Distinguished Flying Cross Society, The Villages Chapter.

The forthcoming pages highlight the faces of several Chapter members and share their stories of heroism and extraordinary achievement in defense of democracy and freedom. Each section provides a window into the life, their career, and the actions they took which led to the awarding of the DFC, as written in their own words.

These are the stories of ordinary Americans who faced the extreme conditions of war to make a difference in the lives of their fellow men and women. We invite you to get to know these Chapter members through their stories, and thank them for their openness, honesty and humility. If you see them on the street, why not take a moment to shake their hand and get to know them.

Whether you consider yourself a Rotor Head, Jet Jock, Prop Jock or just a lover of military aviation and history, we think you will appreciate this book. We hope you enjoy reading their stories, as much as we enjoyed collecting them.

Michael A Gulick
President
The Villages Chapter
The Distinguished Flying Cross Society

ACKNOWLEDGEMENTS

We would like to thank Don Finch for his invaluable assistance in reformatting the book into the proper software version, electronic font size/style to meet Amazon and CreateSpace requirements.

Don also created many of the graphics that grace the book, the section dividers, and arranged the images in various chapters.

We would like to thank Nancy Finch for retyping portions of some chapters.

We acknowledge that many of our contemporaries returned from their war to an America that did not welcome them home, which was done for our heroes who returned from WW II. Readers will note some of us came back to notices that we were to wear civilian clothes, try to be invisible, be aware that protesters would spit at us and throw things at us, so we were asked to try NOT to look like we were military…Returning to the USA in 1971 was not much fun. Perhaps for that reason, we are taking great care to ensure our warriors returning from the Middle East are welcomed with open arms.

Each of the chapters was reviewed by at least two people; we would like to thank Tom Brigham and Tom Branham for their insight on chapters dealing with the early days in Vietnam; Bobby Beeks for his thoughts from a perspective of *tabla raza* knowledge about the military and flying; Betty Finch for her nurse's insights on Medevacs; each of the authors that helped edit other chapters.

Barney Barnhart, Rich Burns and David Lehtonen sat with me for hours brainstorming the process, flow of the book, then each of them spent at least one day with me reviewing chapters. Wayne Lotsberg was quick to respond to my every request for ideas and help.

Michael Gulick gave me broad guidance and let me organize the project my own way.

HIGH FLIGHT

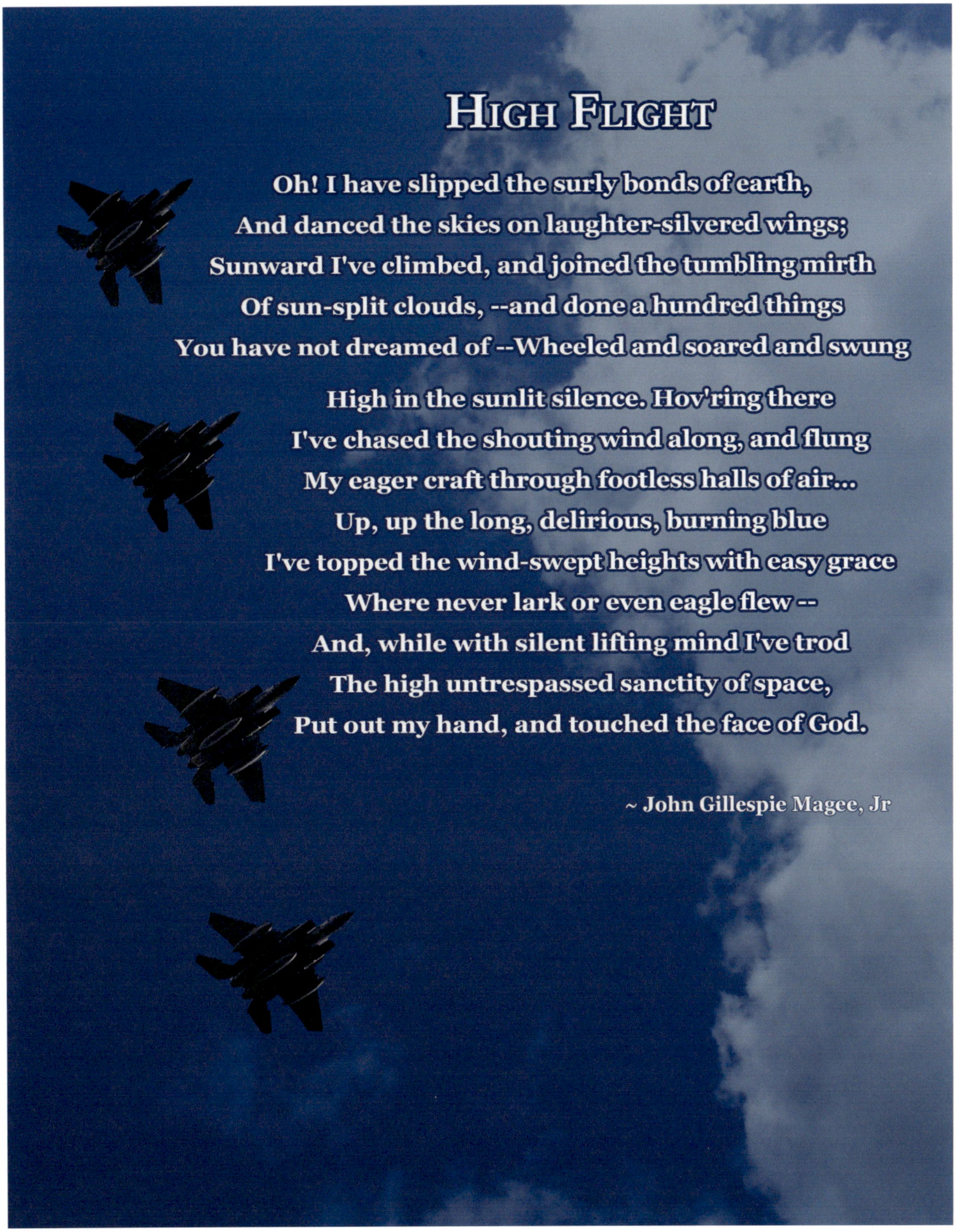
HIGH FLIGHT
Oh! I have slipped the surly bonds of earth,
And danced the skies on laughter-silvered wings;
Sunward I've climbed, and joined the tumbling mirth
Of sun-split clouds, --and done a hundred things
You have not dreamed of --Wheeled and soared and swung
High in the sunlit silence. Hov'ring there
I've chased the shouting wind along, and flung
My eager craft through footless halls of air...
Up, up the long, delirious, burning blue
I've topped the wind-swept heights with easy grace
Where never lark or even eagle flew --
And, while with silent lifting mind I've trod
The high untrespassed sanctity of space,
Put out my hand, and touched the face of God.
~ John Gillespie Magee, Jr

INTRODUCTION

The **Distinguished Flying Cross** is a military decoration awarded to any officer or enlisted member of the United States Armed Forces who distinguishes himself or herself in support of operations by "heroism or extraordinary achievement while participating in an aerial flight."

The men chronicled in this book earned the Distinguished Flying Cross, in many cases earned that award repeatedly without recognition, until a ground or air unit commander noticed their action, found time to write and submitted them for the award. Those like Tom Rozga who earned three DFCs in two wars and Rudy Martin who flew into the face of enemy Anti-air and air-to-air fire repeatedly over an extended period of time were truly heroic. It takes a lot of nerve to fly into a place where you know you will be shot at and you could potentially be killed, yet you flew on into the swirling maelstrom of enemy fire.

There is disparity between the length and level of detail provided in each Chapter. Rightly so! Each chapter was written by the man who flew it, experienced it and survived it. These are firsthand accounts. There are differences in how they write and how they describe events. They had to make instantaneous decisions which would affect the lives of other men, and took responsibility for their actions even when they were well beyond exhaustion after flying as many as 25 or 27 hours in a row while under hostile fire, with adrenalin pumping, and generally without food.

Those who received a DFC are not generally boastful but they know in their hearts they have earned bragging rights, for they have been at the limits of human endurance, limits of performance for their aircraft, and made a difference. Many risked their own lives to save others. In some cases, the performance of their mission ensured peace for another day, in some cases their actions shortened the war and saved lives.

The following chapters are but a snippet of the lives of ordinary men placed in the difficult circumstance of war and made a difference for the United States of America.

Their stories do not end with the award of a Distinguished Flying Cross. Readers will note each of these men went on to live normal, productive, prosperous lives, in a variety of occupations. They have made a difference.

It was my privilege and a distinct honor to listen to these men, help them articulate what they went through, in details they had confided to few other people. I have been awed and a few times in tears as I **listened between the lines** like when Ken Taylor said "we put out the fire" and continued on the mission. Few of us can comprehend the dread that filled the cockpit along with that smoke. Pilots do not like it when their aircraft catches fire…

It is my hope the reader will gain an understanding of what these men endured to be recognized with a Distinguished Flying Cross. Also, I hope you will better understand how some of us became a little quirky in our old age.

"Our debt to the heroic men and valiant women in the service of our country can never be repaid. They have earned our undying gratitude. America will never forget their sacrifices."

President Harry S Truman – National World War II Memorial

FLIGHT FOREVER

SECTION 1: WORLD WAR II

The Second World War was a global war that lasted from 1939 to 1945, although related conflicts began earlier. It involved the vast majority of the world's countries which included all of the great powers and eventually forming two opposing military alliances: The Allies and the Axis. It was the most widespread war in history, and directly involved more than 100 million people from over 30 countries. In a state of total war, the major participants threw their full economic, industrial, and scientific capabilities behind the war effort, erasing the distinction between civilian and military resources.

World War II was the deadliest conflict in human history, marked by 50 million to 85 million fatalities, most of which were civilians in the Soviet Union and China. It included massacres, the deliberate genocide of the Holocaust, strategic bombing, starvation, disease and the first use of nuclear weapons in history.

Source: Wikipedia

Fighter Pilots

Pilots are people who drive airplanes for other people who can't fly. Passengers are people who say they fly, but really just ride.

Lockheed P-38 Lightning

Fighter Pilots are steely eyed, weapons systems managers who kill bad people and break things. However, they can also be very charming and personable. The average fighter pilot, despite sometimes having a swaggering exterior, is very much capable of such feelings as love, affection, intimacy and caring. (However, these feelings don't involve anyone else.)

There are certain aircraft sounds that can only be heard at night and over the ocean. Most of them are scary.

1

Jack H. Hallett
Captain
United States Army Air Corps

World War II, European Theater (1944 and 1945)

Very proud of his flight cadet's uniform

Jack Hallett, was born in Boston, Massachusetts, 17 Nov 1920. Jack's Father was an engineering graduate of MIT who worked as an engineer in various locations near Boston. His father was brilliant. At one point he became the Chief Engineer for the state of Massachusetts. Jack had an ersatz uncle, who had flown in WW I. He discussed his experiences in WW I and had told Jack several times, "If you have to serve during a war… fly."

From an early age Jack wanted to fly, more importantly "I wanted to be a fighter pilot because that's where the excitement was."

On 7 Dec 1941 he was at a hockey game. He heard several announcements over the public address system: "All military personnel are to report to the nearest Military base for further instructions". When he got home, he found out why: the Japanese had attacked US naval forces at Pearl Harbor, Hawaii!

"That evening I discussed it with my family... I told my parents that I 'had to go, and I didn't want to carry a rife in the mud; but that I wanted to fly. -- Dead silence!

"There was an awkward silence, they looked at me like someone had slapped them". Jack's older brother had died in a military aircraft crash. "They ***knew*** I had to go into the service, and finally said, OK."

Everyone knew things could get ugly; but Jack had made up his mind he wanted to join up so he could become a fighter pilot. The next day, He went down to the recruiting office. "It took 3 - 4 hours just to get in the Armory building. There were long lines, everyone wanted to serve" (Jack says, "I would like to think people would want to sign up like that today, if there was a War.")

"I asked to be an aviation cadet; told them I wanted to be a fighter pilot." Jack had no college but would be allowed to take an exam to demonstrate two-year college equivalency, including ability to solve problems, write, and do math. He passed those tests well enough and passed all the physical requirements including depth perception, 20:20 eye sight, color vision and hand-eye coordination. "I was allowed to join the cadet corps. I was signed up and getting about $21.00 per month as a Private; but I had to wait for a flight class opening, so I found some work."

"I hope the IRS doesn't come after me for working for both the Army Air Corps and the US Navy at the same time"; After I signed up with the Army I was told to wait until a class date was available. I stayed around the house, but figured I should do something constructive. At that time "Dad" was the chief engineer for the Hingham Ammunition Depot construction site. They were building large concrete Igloos shaped like Nissen huts. These igloos were for storage of ammunition for the Navy and the building specs called for an inspector to watch each pour and sign off on it . That was my job even though I knew nothing about concrete work. At that time all the concrete workers were Italian and only the Foreman spoke much English. Dad jokingly told me not to make them angry or I might disappear in one of the forms.

Pay as a Private was $21.00 per month; Jack boosted his income to make as much as $80.00 by working seven days a week for the US Navy as a concrete inspector for construction of the Ammunition depot in Hingham, MA. In other words, he was paid by the Army Air Force $21.00 per month as a Private, and quite a bit more inspecting concrete for the Navy. (Jack was an early double dipper.)

"In March 1942 I was called and a group of us from the Boston area were put on a train to Maxwell Field near Montgomery Alabama, for 'classification' and preflight training. Here we all got military training, weather, Morse code, Theory of Flight, rigorous physical training, calisthenics and Drill and Ceremonies. This is where, through various tests, decisions were made to place us in Pilot, Bombardier or Navigator training. I made Pilot! At Maxwell there were quite a few eliminations for multiple reasons.

"In June 1942 I was assigned to Lafayette LA for primary flight training. I began flight training in the PT-17 Stearman; the fuel gauge was a glass vial with a cork in it, placed in the top wing. You could only read it when you were in straight and level flight. Additionally, when you were inverted, the engine could conk out. There have been a lot of improvements in aircraft over the years."

The PT-17 Stearman

"There have been a lot of improvements in aircraft over the years."

Vultee BT-13 Valiant

North American Aviation T-6 Texan

"In flight school we began flying training in the PT-17 Stearman at Lafayette LA, I soloed with 8 hours in the Stearman in July 1942, completed primary flight training with 60 hrs. From Lafayette, I went to Greenville MS for basic flight school, flew the BT-13 which gave us some proficiency in night and instrument flying and another 70 hours. Then we went to different bases for advanced schools. Those who were going to be multi engine pilots went to multi-engine flight training schools; I went to Craig field near Selma, Alabama for single engine training, graduated Jan 14 1943. I got another 70 hours. including instrument and night flying in the AT-6. On the day we were commissioned all of us were behind in the number of flight hours required to receive our wings; poor weather conditions had grounded our aircraft several times during the flight training program".

"Prior to graduation, about 8 of us were allowed to transition to P-40's. The transition included a 'blindfolded cockpit check'—with a scarf tied over his eyes he had to show proficiency in locating controls and gages, flaps, throttle, mixture control, landing gear, brakes etc." When he took off, he climbed to altitude, practiced stalls, turns, climbs and dives, then flew around the traffic pattern for a landing. The first landing taught Jack just how much different the P-40 handled compared to the training aircraft they had been flying. Jack says, "It was sort of a 'white knuckle flight.' I was drenched in sweat when I finally landed." It was OK. I was in a fighter and would soon be a fighter pilot, but; when I finished my flight training, they assigned me to the flight training school teaching cadets." Because he had flown P-40's, Jack says, "I went to the Colonel and told him there had been a mistake on my orders". When he asked what the mistake was, "I told him I had signed up to be a fighter pilot not a flight instructor." Jack told the commander that he had qualified in P-40's and wanted to go into combat as a fighter pilot… The Colonel scoffed and told Jack, "Goodbye, get out of here son, go on about your day." (Jack was a little miffed. He was now a second lieutenant with several days in grade, and over 21 years old. He thought he deserved more respect than to be called 'son'.)

P-40 Warhawk

Jack wanted to be a fighter pilot not an instructor. At least he was flying, not slogging through the mud with a rifle as an Infantryman. He became an instructor in the BT-13's in the newly-opened Army Air Field, Courtland, Alabama where he met his future wife, Martha Jane Adcock, on a blind date. "I was doing everything I could to get into combat as a fighter pilot". Jack's then fiancé didn't want him to go. "She wanted to get married before I went overseas. I was afraid she would get pregnant; and then if I didn't make it back, she would be a 19-year-old widow with a child. We didn't have sex before marriage in those days. I told her we would have to wait to be married when I got back from the war.

Jack was proficient enough and feisty enough to get his way. While he was instructing for the Eastern Training Command in Courtland AL, he flew numerous other airplanes (e.g. AT-10's AT-11's, PQ-78, AT-6 and UC-78). In early 1944 orders came through for Jack and four classmates to go to the West Coast: They were reassigned to the West Coast to complete Fighter training in preparation for deployment to Europe!

When he got to the West coast, he reported to Moses Lake, Washington, for training on the P-39; then to Ontario, California where was able to transition to P-38 Lightnings. The transition consisted of ground school, several hours of briefings and about 80 hours of flying.

The P-38 had a lot of power, with light wing loading[1],(making it very responsive). They flew at Moses lake, Washington, and Muroc [Muroc Maru, officially AAF Temporary Building (Target) T-799, was a replica of a Japanese Takao-class cruiser constructed on the floor of Rogers Dry Lake in southern California during World War II. Used to train bomber pilots and bombardiers in techniques for attacking ... Army Air Force pilots assigned to train at the nearby Muroc Army Air Field.] now part of Edwards Air base, California and flew out over the Pacific where they practiced some aerial gunnery shooting at sleeves towed by a B-26. Learning how to aim the guns of the P-38 Lightning was easy. The guns were on the centerline, in front of your legs. The distance to target and airspeed when you fired the guns did not matter much. The nose mounted guns on the P-38 were four 50 caliber machine guns and a 20 MM cannon."

Finally, in 1944, Jack got an assignment to Europe. "I took a train to New York, then sailed on the Queen Elizabeth I to England, landed in Firth of Clyde, Scotland.

"I was assigned to the 367th Fighter Group, 364th fighter Squadron. The 9th Air Force was responsible for '***tactical*** flying': close air support, tactical bombing of targets, and bomber escort, in short fighter pilot flying; the 8th Air Force was responsible for ***strategic*** flying: long range strategic bombing and bomber escort." When Jack completed his flight training there were only three P-38 groups in the 9th US Army Air Force.

D-Day Invasion

Jack flew air cover in a P-38 over the D-Day invasion of Normandy (6 June 1944). "I had maybe 90 hours in the P-38 when they sent me into battle with it" He spent most of the Normandy invasion watching for enemy airplanes, but the expected enemy fighters didn't show up. (As Jack says, with a smile and a twinkle in his eye, "The Jerry's found out I was there in a P-38 and they stayed away.") "We shot at targets of opportunity, transports, cars, even bicycles, or whatever target the forward observers called to our attention." After D day we set up air strips and maintenance areas in cow pastures or whatever safe areas we could find. Those became our temporary airfields and maintenance areas.

"I loved being a fighter pilot"

---

[1] Wing loading: Pilot talk for agility, responsiveness, & sensitivity of controls: also, how heavy the aircraft "feels" in flight: Heavy wing loading would imply the A/C seemed Very heavy, or sluggish; versus light & nimble;

Battle of the Bulge

December 15th or 16th, 1944[2]. The Germans mounted an all-out attack in what they thought was a weak area in the Allied lines. This was in the Ardennes forest in Belgium. The purpose was to split the Allied forces and push all the way to Antwerp. The weather was bitter cold, low visibility and snowing. No way the fighters could be used against the enemy.

I don't recall how long we were grounded but it seemed like weeks. Knowing how desperate our GIs were for help, if there was a small break in the clouds, somebody would try to go out individually only to abort because of the weather again. When the skies finally cleared, we loaded up and sent a lot of the enemy to Valhalla along with their equipment, tanks, trucks, etc. It was a field day for us. We did lose a couple of aircraft to ground fire over the next few weeks. However, our losses were slight. Our poor GIs on the ground below went through hell through this period but somehow managed to hold their ground in spite of the weather, the overpowering forces, and the continuing heavy fire of all kinds. ("I wonder if I would have had the courage to go through what they did on the ground and keep fighting as they did".)

The 9th Air Force fighter groups continually moved bases as our ground forces liberated new areas. During the winter of 1944-45, we took over what had been a German airfield and, as I remember, it had a paved runway. There were also numerous wrecked German airplanes on the perimeter that had been bombed and strafed by our Allied aircraft.

The airfield also had barracks that were like a center-aisle motel, where we stayed. The rooms were small, four of us to a room, warm with a wood burning stove. The outside walls were thick concrete and one end had been bombed out but that end was boarded up. We were lucky not to be in tents. That base was at a town called Juvencourt, east northeast of Rheims.

"I was a captain at the time I earned my DFC—the order states '1st lieutenant'—the order is wrong. I accumulated a total of about 300 hours in the P-38. One thing we seem to have forgotten is how hard the crews worked. I had a crew chief, assistant crew chief and an armorer. When we pilots got home safe at the end of each day the crews went out and worked over the aircraft to patch holes made by enemy fire; crews serviced the engines, hydraulics, electronics, radios and armorers worked on our guns and re-armed the aircraft. All was done to ensure the aircraft were ready for action by the next morning—they often worked through the night in frigid cold or rain. I don't recall ever having a hangar. (We may have had hangars in England before D- Day.) The crews made sure the aircraft were safe to fly; they were largely

[2] The Battle of the Bulge: Lasting six brutal weeks, from December 16, 1944, to January 25, 1945, the assault, also called the Battle of the Ardennes, took place during frigid weather conditions, with some 30 German divisions attacking battle-fatigued American troops across 85 miles of the densely wooded Ardennes Forest.

responsible for our safety while flying. They took pride in their work and in the aircraft and pilots they served. If I made a bad landing the crew would give me hell. These unsung heroes received very little recognition for their extraordinary work. "

Bill Mason (left) and Jack pause between missions

We moved a lot as the Allies pushed relentlessly on to Berlin. Bill Mason (above left) and Jack were good friends; they flew about 60% of their missions at the same time but in separate aircraft. They got to know what each other were thinking as soon as situations developed, they anticipated each other's actions – made a very effective team. "Bill Mason and I never flew wingman for each other but normally we flew the same missions and we roomed together."

Jack talks about how pilots identified friendly versus enemy vehicles on the ground. Friendly vehicles had a set of brightly colored panels placed on the top of each vehicle with various colors depending upon the day and the mission. The fighter pilots had to identify those panels from two thousand feet or more, and they were coded. The codes changed day to day, and in heavy conflict they often changed two or three times in a day. As the pilots got their morning briefings before takeoff, they were told what color of panels would identify friendly vehicles.

In December 1944 Jack was attributed with an enemy kill. He was flying escort with a flight of British Lancaster bombers on their way to Berlin. He periodically checked his rear-view mirror. He spotted four Messerschmidt M-109's on his tail. He tried to lose them unsuccessfully a couple of times; but their gunfire was getting closer; he remembered a radical maneuver his flight instructor had mentioned to escape enemy on your tail: the 'aggravated spin.' He pushed one throttle full forward, idled the other engine, pulled back on the stick, slapped opposite rudder hard ...; "I don't really know what happened to the aircraft, it spun, climbed, yawed, rolled, finally stabilized. I don't really know what we did, but, the M-109s were nowhere to be seen." [Can you, for a moment, try to visualize this from the enemy fighter pilot's point of view: Suddenly the predictable flight path of an aircraft in front of them does a 'WHAZZAT' maneuver right in front of them, makes a flat spin to end up pointing at them and climbing... How could a pilot or an airplane do such a thing??] "It's a wonder I didn't break my damned neck.", Jack says. The enemy pilots probably wondered, "What the hell was that?"

P-38 Lightning

P-38 Lightning

On several occasions as his squadron attacked German trains, they watched with some excitement as a railroad car would explode into a huge ball of fire and debris. They had hit a rail car full of German ammunition! Sometimes his aircraft would fly through the fireball and he could feel the heat of the explosion. ("It may not seem apropos but that part of flying in the war was FUN." Oh, looky here! D'ya see that! Woo Hoo!")

Fighter Pilot with 1 Enemy shoot-down – Not quite an Ace!

On an escort mission in Dec 1944, Jack discovered enemy aircraft closing in on his group. He turned to attack one aircraft and got into a dogfight. He was able to get above the enemy craft, aimed and fired. The enemy pilot dove to avoid his fire, then inverted and went into low clouds, as Jack followed right side up. The enemy pilot discovered the low clouds were more like a fog bank just a couple of hundred feet above the ground. He was unable to pull up in time and crashed into the ground. Jack detected the fireball and was able to pull up in time to avoid hitting the trees. Jack was able to pull up in time due to his quick reflexes and the superior maneuverability of his P-38. Did he hit the pilot injuring him, thus causing the crash, or was there a catastrophic hit to the aircraft? No one is sure on the exact details but Jack got credit for one enemy plane shot down: 4 less than the 5 'kills' required to become a fighter "Ace".

Jack was shot down two times, once in a P-38, once in a P-47. That was not as much fun as blowing up a train car full of explosives. One of those times his P-38 aircraft was hit by enemy gunfire and he knew he had to land soon. He landed at their field which was "paved" with perforated steel planking (PSP). He thought the landing was going well, put his flaps down for landing, put the wheels down got three green lights indicating the landing gear were down and locked, then made his final approach. When he touched down the airplane lurched and hit the ground hard, cartwheeled and banged him around in the cockpit "What a surprise." His beautiful P-38 was a mess. Eventually they discovered that although the left gear was down there was no wheel at the end of the strut; possibly it had been shot off or fell off while he was flying to safety. Looking at the aircraft, no one could understand how it could have happened. "It just wasn't possible. That should not have happened, but did."

Flying Scared

"There were times when we all really got scared, but wouldn't admit it when we got back on the ground. One mission in a P-38 I was positive at the time I would not survive. The mission was to bomb and strafe a German marshalling yard at the junction of the Rhine and Mosel rivers. The yard was on the East side of the Mosel river near the mouth in a valley with steep hills on each side of our target. I don't remember if there were 12 or 16 of us, but I was leading the last flight of four aircraft. As the others went down to dive bomb and make their strafing runs, we stayed up for top cover until they were done, then it was our turn. We watched as the guys dove down and drew enemy fire. The hillsides lit up with cross fire, 88's, 40 mm's, 20mm, and .50 caliber machine gun fire. You could tell by the color of the tracers and smoke the caliber of the various guns. I remember thinking, 'If all the lead were to stop and stay in midair right where it was, and you could wiggle at low airspeed, there was no way to get through all those bullets without getting knocked down'. Our turn came soon enough. Because the other guys did, the four of us went down bombing and strafing knowing we would face the same intense barrage. We left a hell of a mess, flew like the rest of the squadron, stayed low on the river and got out of there. Somehow we didn't lose anyone. A few holes in our aircraft but everyone got home safe. There may have been some changes of underwear but we were safe."

In 1945 the War in the Pacific was in need of the P-38's so Jack's unit transitioned to P-47's. "We received three P-47s to assist us in the transition. On the first day one crashed after the pilot bailed out so our flight group of more than 100 pilots learned to fly P-47's on just two aircraft. It took about an hour or two to transition from the P-38 to the P-47. Jack had about 55 minutes in the P-47 when he flew it in combat. We were so accustomed to the P-38's nimble flight characteristics, aerobatic agility, power, and its responsiveness. It was a delight to fly. It could fly you back to base on one engine. The P-47 on the other hand had only one engine, heavy wing-loading with four .50 Caliber machine guns in each wing and very sluggish response to pilot inputs on the controls especially when compared to the P-38. In the P-38 when you pulled the nose up it darted up quickly; when you pulled back on the stick in the P-47 the attitude changed but the altitude lagged behind. That difference caused several crashes as pilots learned the P-47 was not as responsive.

Once we got used to the P-47 it was probably a better airplane for our missions—close ground support. The maintenance on the P-47 was less extensive than the P-38. The P-47 was a great airplane, less expensive, easier to produce and, most importantly, it could take a hell of a beating, still bring you home safe. The Pratt and Whitney R-2800 was a powerful and reliable engine.

P-47 Thunderbolt – "Frigid Midgit"

This picture of the P-47 is an exact replica of the one Jack flew; The logo on the side was mis-spelled but the paint job was so beautiful we left it as it was.

Jack sitting on wing, relaxing

The Distinguished Flying Cross

Jack received his DFC, for action on 17 March '45; It was a cold rainy day, with a low ceiling, and low visibility. It was difficult to navigate to the target. He was flying a P-47 Thunderbolt. He was unable to follow the railroad tracks, and had to fly time-distance-and-heading or "dead reckoning" to reach the target near Birkenfeld, Germany. His target was an area of Germany between Luxembourg and Mainz consisting of fields, a few houses and assembled enemy troops. Their target was to attack any transports including tanks, trucks, automobiles, and any fixed facilities with enemy activity. They were opposed by severe machine gun and ground fire: Quad 20's.

By Bundesarchiv, Bild 183-J08339 / Schwahn / CC-BY-SA 3.0, CC BY-SA 3.0 de, https://commons.wikimedia.org/w/index.php?curid=5434325

Flakvierling Quadmount 20mm

The orders awarding Jack Hallett a Distinguished Fling Cross are quoted below:

Headquarters Ninth Air Force
General Order Number 119, 29 June 1945:

"The Distinguished Flying Cross is awarded to:

JACK H. HALLETT, 0-795591, 1st LT, Air Corps, 367th Fighter Group ... For extraordinary achievement in aerial flight against the enemy. On 17 March 1945, LT HALLETT, leading his squadron on a patrol mission without bombs in the vicinity of Birdenfeld, Germany distinguished himself by outstanding aerial skill and leadership. In spite of adverse weather conditions and severe enemy fire, he located and seriously damaged or destroyed over one hundred enemy transports, thereby contributing toward the dispersal of enemy forces in this area. The courage and gallant devotion to duty reflect great credit upon himself and the Army Air Forces. Entered service from Massachusetts."

Jack served 4 years during World War II. He did not serve a full military career, but got out of the service on 14 February 1946. He had flown 104 missions in combat, had earned 13 Air Medals and a Distinguished Flying Cross. He had "shot down" one German fighter pilot in a dog fight and had helped the Allies achieve victory in the Greatest War. He also helped forge the legacy of the Greatest Generation. At 99 years of age he still gives presentations on his perspective of the war and he has a host of good friends around the USA.

In 1951 he and his wife moved to Florida. Initially he taught flying at Bartow air base near Lakeland Florida. He was a flight instructor from 1951 to 1960. He was teaching pilots who would then fly in Korea. Jack was a Flight Commander from 1955 to 1960 responsible for 80 to 100 aircraft and numerous students. The students had T–6, T–34, T–28, and T–37 trainers. "We had an extraordinary safety record. At one point the base went 90 days without any safety incidents on the ground or in the air". The base closed in December 1960.

Jack opened an irrigation business. He developed a lucrative contract with Disney. The irrigation business was doing well but he was not paying enough attention to what was going on financially. "It was my own fault. Pretty soon we owed the IRS a lot of money, and had to declare bankruptcy."

" I had no more income so I went to work seven days a week doing whatever work I could find to keep food on the table and make our payments. Eventually we were able to get ahead, set some money aside, and made a couple of good investments. We were able to pay off the house, and life was good in Florida. My wife, Martha, developed heart problems, she passed away in 1986. Our community is very nice, I have friends here, the home is paid for, and I am happy with my life. I have met a number of very nice people and have good friends all around the country."

As recently as 2019 Jack has flown in acrobatic aircraft and still loves the sounds sights and smells of flying; In 2016 he was allowed to fly the simulator for the F-35 Lightning *II* – yes, the successor to the famous P-38 lightning he had flown in WW II. He even made a carrier landing with the flight simulator. Jack was also allowed to fly a P-51 in 2019. Thus, he was able to fly all of the aircraft in the picture below:

**Jack Hallett has flown each of the WW II fighter aircraft pictured here.
The P-38, P-40 and P-47 in combat, the P-51 mustang (top) he was able to fly in 2019 as a 98 year old veteran fighter pilot.
Missing in the above picture of fighter aircraft Jack flew is the R-39 Bell Air Cobra.**

Jack Hallett at 99 years old

"If I am to be remembered for anything, I want it to be that I was a fighter pilot! I loved being a fighter pilot."

SECTION 2: KOREAN CONFLICT

The Korean War, 25 June 1950 – 27 July 1953, began when North Korea invaded South Korea. The United Nations, with the United States as the principal force, came to the aid of South Korea. China came to the aid of North Korea, and the Soviet Union gave some assistance.

In August 1945, the Soviet Union declared war on Japan. As a result of an agreement with the United States, Korea was split into two regions, with separate governments. Both claimed to be legitimate. The conflict escalated into open warfare when North Korean forces moved into the south on 25 June 1950.

After the first two months of war, South Korean and U.S. forces were on the point of defeat, forced back to a small area in the south known as the Pusan Perimeter. In September 1950, an amphibious United Nations (UN) counter-offensive was launched at Incheon, and cut off many North Korean troops. UN forces rapidly approached the Yalu River, border with China, but in October 1950, mass Chinese forces crossed the Yalu and entered the war. The surprise Chinese intervention triggered a retreat of UN forces which continued until mid-1951.

Seoul changed hands four times. The last two years of fighting became a war of attrition. The war in the air, however, was never a stalemate. North Korea was subject to a massive bombing campaign. Jet fighters confronted each other in air-to-air combat for the first time in history.

The fighting ended 27 July 1953 with an armistice. The agreement created the Korean Demilitarized Zone (DMZ) to separate North and South Korea, and allowed the return of prisoners. However, no peace treaty has been signed. The two Koreas are technically still at war.

Source: Wikipedia

F-86 Sabre

He who demands everything that his aircraft can give him is a pilot; he who demands one iota more is a fool.

North American F-86 Sabre

The North American F-86 Sabre, sometimes called the Sabrejet, is a transonic jet fighter aircraft. Produced by North American Aviation, the Sabre is best known as the United States' first swept-wing fighter that could counter the swept-wing Soviet MiG-15 in high-speed dogfights in the skies of the Korean War (1950–1953), fighting some of the earliest jet-to-jet battles in history. Considered one of the best and most important fighter aircraft in that war, the F-86 is also rated highly in comparison with fighters of other eras.

2

David E. Lehtonen
Lieutenant Colonel
United States Air Force/CAP

Korea & Cold War

Lieutenant Colonel David E. Lehtonen

Mr David Lehtonen, a 1951 graduate of Lunenburg High School, joined the Massachusetts National Guard in January 1951 and trained as a radio operator with the 181st Infantry Regiment, 26th Infantry (Yankee) Division, which met at the Fitchburg MA Armory. He was attached to Headquarters and Headquarters Company where he learned the manual of arms, close order drill, and how to fire the M1 Garand rifle, a gas operated weapon weighing about 9.5 pounds. Before summer camp at Ft. Drum, NY, he trained as a radio operator under the guidance of Sgt George Aho, a 1941 Lunenburg High graduate who served during WWII. Sgt Aho taught him how to set up and operate a jeep mounted HF radio and to establish communication with other National Guard units around the state. Sgt Aho and PFC Lehtonen would go to the top of nearby Rollstone Hill and set up a vertical antenna during drill night, operate the radio net, and after about an hour they would dismantle the antenna and return to the Armory. Continuing his infantry training, Pvt Lehtonen was issued a new M1 rifle that was manufactured during WWII and protected in cosmoline over all the metal parts. Cleaning this rifle before use was quite a task because the cosmoline is a tough impervious material, resistant to most solvents. However, lye soap and water with a measure of elbow grease did the job. Once the rifle was cleaned, he was shown how to bore sight the weapon to adjust the sights to prepare for live firing at Ft. Devens in Ayer, MA that was conducted later during weekend drills.

With the previous as background, during the summer of 1951 I was assigned to training at Ft. Drum, NY for what amounted to an accelerated basic training program. During this training, I received plenty of time on the rifle range and parade ground as well as live fire obstacle training under barbed wire with 30 caliber machine gun fire over our heads. I managed to shoot expert marksman with the M1 on the range and remembered not to stand up during both the 30-caliber live fire obstacle training. Also, I was given live fire familiarization with the 50-caliber machine gun and the 30-caliber machine gun as well as having bayonet practice and of course, close order drill. This training was intense because the Korean War was just starting up and the 26th Division was likely to be called up. The last time the Yankee Division was federalized was during WWI where the division served in the trenches in France, taking many casualties. Looking around the company, I thought that if the 26th Division was called up and sent into battle, we were too poorly trained and ill prepared to fight, so the casualty rate could potentially be high. It was at this point that I decided to enlist in the U.S. Air Force and so in January 1952, I was honorably discharged from the Massachusetts National Guard and sent to Sampson AFB, NY for basic Air Force training. As it turned out, the Yankee Division was not called up for active duty but I felt fortunate to have had the infantry training and to serve on active duty as a radio operator in a B-26, flying 50 combat missions over North Korea earning the Distinguished Flying Cross and the Air Medal with 2 Oak Leaf Clusters.

U.S. Air Force Service, Korean War

Sampson AFB, a former US Navy training base during WWII is located on Lake Geneva, one of the Finger Lakes in upstate NY. January 1952 was a transition period when the

Air Force was moving from Army Air Corps khaki into the New US Air Force blue uniform. Because of Korean War shortages, not all clothing items were available. I was issued a field jacket without a cold weather liner and but for the long johns we wore during that bitter cold eight weeks, our basic training flight personnel could have been worse off than we were. As it was, over half of the base came down with pneumonia, filling the base hospital to overflowing. I was lucky because when I got sick, I was able to stay in the barracks pumped full of antibiotics while my buddies brought me my meals. Had I been admitted to the hospital, I would have been put back a week, so as it turned out, I was able to finish my basic training on schedule. Near the end of this training, my flight was able to have a weekend family visit. My family drove out to New York from Lunenburg in a snowstorm to visit me. The visit was much appreciated, giving me a break in the intense training.

Sampson Air Force Base Memorial

The Sampson AFB Memorial, shown above, honors those airmen who trained there during the Korean War. The base closed in 1956 and is now a NY State Park.

After completing basic training, a lot of the people in the flight were sent to cook school or air police because that was the need at the time. Because of my earlier training, I was fortunate to be assigned to Keesler AFB, MS because I held a radio operator Military Operations Specialty (MOS) while in the National Guard. So, in March 1952, I got on a train for the Radio Operator General Course 29351, Class 24032-C at Biloxi, MS. This course concentrated on teaching proficiency in Morse code along with the various receiving and transmitting equipment used at the time. The airborne equipment we trained on was the ARC-13 transmitter and the BC-348 receiver while a BC-610 transmitter was used for ground installations. Electronic theory and antenna theory were also part of the course that took eight months to complete. Upon graduation, the Air Force gave each class a choice of assignments and of the 36 airmen in this class, all but two asked for Europe. Those two wanted to go to Japan and Korea to "get into it." Well, you can imagine what happened. The two who asked for Japan and Korea were sent to Europe and the balance of the class was assigned to various squadrons in Japan and Korea. Reverse psychology? I was assigned to the 5th Air Force, 67th Tactical Reconnaissance Wing located at Kimpo AFB just north of the capitol Seoul. But before leaving to the Far East Air Force, everyone got a 30 day leave which I spent at home in Lunenburg, Massachusetts with family.

At this point a word about the Korean War is in order. According to information from *The USAF in Korea: A Chronology 1950-1953*, Air University Press, 2000, the North Korean army invaded South Korea on June 25, 1950, driving the South Korean army and a small US force attached to the Republic of Korea (ROK) army south to Pusan in what can best be described as an orderly retreat. The North Korean government and the Soviet Union believed that the UN would do no more than complain loudly at this invasion. But the UN Security Council on the same day unanimously called upon all member nations "to render every assistance…" to the UN to fight this aggression. President Harry S Truman in his role as U.S. Commander-in-chief, issued an order on June 27th for American air and naval forces to support South Korea, choosing General Douglas MacArthur to lead the joint UN forces. In September 1950, General MacArthur's Marines successfully invaded the Korea's western shore at Inchon, cutting the North Korean supply lines and forcing their rapid retreat from the south. The combined Army, Marines and ROK forces pushed north to the Choshin Reservoir by November when the Chinese invaded across the Manchurian-Korean border in overwhelming numbers, driving the UN forces back to the pre-invasion 38th parallel. General MacArthur was recalled to the US by President Truman on April 11, 1951 because of a differing opinion on how the war should be conducted. He was replaced by General Matthew Ridgeway who began the push north again. But in June, the Russian delegate to the UN proposed that negotiations begin for a cease-fire, a process that dragged on for more than two years, during which time fighting continued in a somewhat stalemated fashion. It is in this context that I arrived in the battle zone in December 1952.

After spending my post technical school leave at home, I had orders to go to Boston to board a chartered DC-3 aircraft to travel to Camp Stoneman in the San Francisco area. This flight took 13 hours and numerous stops were made along the way to pick up other people enroute to the debarkation point. Camp Stoneman was home for a little more than a week when we boarded a ferry to transport us to the San Francisco docks where we would board a Liberty troopship that would be our home for the next two weeks. I recall standing in the ranks on the San Francisco dock very early in the morning, cold and miserable, and the Salvation Army and Red Cross came to distribute hot coffee and donuts, one group on each end of the ranks. The Red Cross was charging for the coffee and donuts while the Salvation Army was giving them away. Life experiences like that seem to be etched indelibly on our minds to the extent that to this day I have little use for the American Red Cross.

The voyage to the Far East was very uncomfortable because the weather off the California coast during the winter is very rough and the air very cold. Add to that the sleeping accommodations were canvas bunks stacked four high with two feet between your bunk and the person above you. And, depending on where your berth was located on the ship, the side to side rolling and the vertical motion could be quite severe. If one was located near the bow, the motion was worse than if you were near the center of the ship. Most people got seasick, some worse than others, depending on where their berth was located. When this sick-to-the-stomach urge happened, one climbed up the companionway, ran to the railing, and threw up over the side. Just make sure when you did that you were near the bow. Otherwise, the downwind people at the rail would catch the spray from the upwind people, resulting in contaminated clothing and making the air in the sleeping quarters quite foul smelling. Another challenge was meal time. Steel tables with a ridge around the edge were bolted to the deck and everyone ate standing up. As the ship rolled from side to side, the metal tray that held your meal had the propensity to slide from one side to the other unless you held on tight. I recall seeing a food tray slide to one side and the person it stopped in front of threw up on the meal. Then when the tray slid back to the first person, and he did the same. One good thing about the trip was that buttermilk was served for lunch and dinner and not many people cared for it so I was fortunate to drink as much as I cared to. The troops disembarked at Yokahama, Japan and those who were bound for Korea were bussed to Tachikawa AFB for airlift to Korea.

The date was December 13, 1952, and this was to be my first flight in a military aircraft, a Fairchild C-119 filled with troops with their gear strapped to the center of the cargo area. Canvas bucket seats lined both sides of the fuselage and we were able to see out the side windows. Unfortunately, the aircraft lost the right engine shortly after takeoff and the flight had to be terminated to return on a single engine. Fortunately, we had enough altitude so were able to make it back OK, except for a brake fire on the right side during landing. We left the aircraft and had to sit on the ramp for a long time while arrangements were made for another aircraft. Our group finally had an uneventful flight to Seoul (K-16) where those with orders for Kimpo AFB (K-14), boarded 6X6 trucks for the short ride north. It was dark when we arrived and we were issued bedding and given a hot meal. No sooner had we got settled in at about 11 p.m., the air raid siren sounded and everyone was told to get up and get into the slit trench just outside the tent. This was the first of many "Bedcheck Charlie" air raids that claimed a few casualties but mostly were designed to harass and keep everyone awake. December 13th was certainly a memorable day! A few days before that, President-elect Dwight D. Eisenhower came to review the Korean situation as he had promised in his presidential campaign. I was sorry not to have seen him.

Life took on a routine that consisted of rising at 0600 hours, cleaning up and making our bed, and having breakfast before going to our duty station. Eight men shared a WWII squad tent that consisted of a tent stretched over a wooden frame with a wooden floor and a kerosene potbelly stove for heat.

Airman Dave Lehtonen, WW II squad tent in Korea

Airmen David Lehtonen and David Gibbs

During the day we had a Korean houseboy who washed our bedclothes and cleaned up. The tent is shown here with David Gibbs, a friend that I have kept in touch with all these years. For his housekeeping duties, we paid Kim a nominal weekly fee in script money. Initially, I was assigned to the 67th Communications Squadron to pass radio traffic as a ground radio operator. This continued for a number of weeks until Communist guerillas broke into the communication compound one night and killed the on-duty crew. I had gone into Seoul (pictured below) that afternoon and evening having swapped my assigned shift with one of the crew that was killed.

On 8-hour pass in Seoul

I'm sorry to say that I don't even remember his name. Shortly after that, a call came down from an adjacent flying squadron, the 6166 Air Weather Reconnaissance Flight, requesting radio operators. I volunteered for one of the two openings and after being interviewed by the squadron commander, Major Laurence Uebel, I was accepted and assigned to a new crew. My aircraft commander's name was Captain Robert Matthews, a recalled P-51 pilot from WWII; the navigator was 2Lt Leonard Hildebrandt from Texas; the weather observer was SSgt Thomas Young from Cape May, NJ; and myself as radio operator. We crewed an RB-26C code named "Blackbird," an aircraft that was used in WWII as well as in Korea - and later in Vietnam as well - obtaining weather data for F-86 air strikes over the Yalu River's MiG alley. A painting of a B-26 434555 is shown in which I flew 4 of my 50 combat missions. My crew position was located aft of the wing and in front of the vertical stabilizer. Access to the radio position was through a small door on the right side of the fuselage.

Douglas RB-26C Invader

My first combat mission was on 23 February 1953, a Charlie II orientation flight in the bucket seat behind the navigator, with another radio operator sending the coded weather information obtained by the weather observer. The next photo shows me with my combat gear standing on the pilot's seat for the picture.

"For great is your love, higher than the heavens; your faithfulness reaches to the skies." (Psalms 108:4-5 NIV).

Airman Lehtonen, dressed in combat gear

There were various parameters we were interested in to obtain weather readings such as altitude above ground, of cloud bases and tops, temperature and pressure at various altitudes, and winds aloft for use in planning effective air strikes against the enemy. There were four different missions: Able, Baker, Charlie, and Easy. The Able mission flew NW to Pyongyang, then NE to Pyorha-ri, a point near the Yalu River, then SE to Mayang-do, an island just off the east coast, then SW to Wonson, then SE to Kangnung, and back to Kimpo. The Baker mission flew west to point Oboe, the island of Paengnyong-do where over-water altitude calibration was made, some 110 miles west of Kimpo AFB, then NW to a point in the Yellow Sea (39 deg N Latitude, 124 deg W Longitude), then SW to another point (38 deg N Latitude, 122 deg W Longitude), then skirting the Chinese mainland to a point (37 deg 15 min N Latitude, 123 deg 30 min W Longitude) and back to Kimpo, all over water. The Charlie mission flew NNW to Sinmak, then NW to Chinnampo, then NE over Pyongyang, then E to Wonsan, SE to Kansong, W to Ch'orwon and SW back to Kimpo. There were four Charlie missions, I through IV, all over the Korean peninsula, all variations of the basic Charlie mission over the Demilitarized Zone (DMZ). The longest of these routes was the Easy mission, which generally was over six hours in length. This route took us west to Point Oboe, then SW to a point in the middle of the Yellow Sea (36 deg N Latitude, 123 deg W Longitude), then SE to a point (30 deg N Latitude, 124 deg W Longitude), less than 100 miles east of the Chinese mainland, then NE back to Kimpo. As this mission had no electronic means of determining location, we relied on dead reckoning with a drift meter. Once I recall that we broke out of the clouds at our southernmost point and found ourselves over some islands off the coast of Shanghai, China where we were warmly welcomed with anti-aircraft fire.

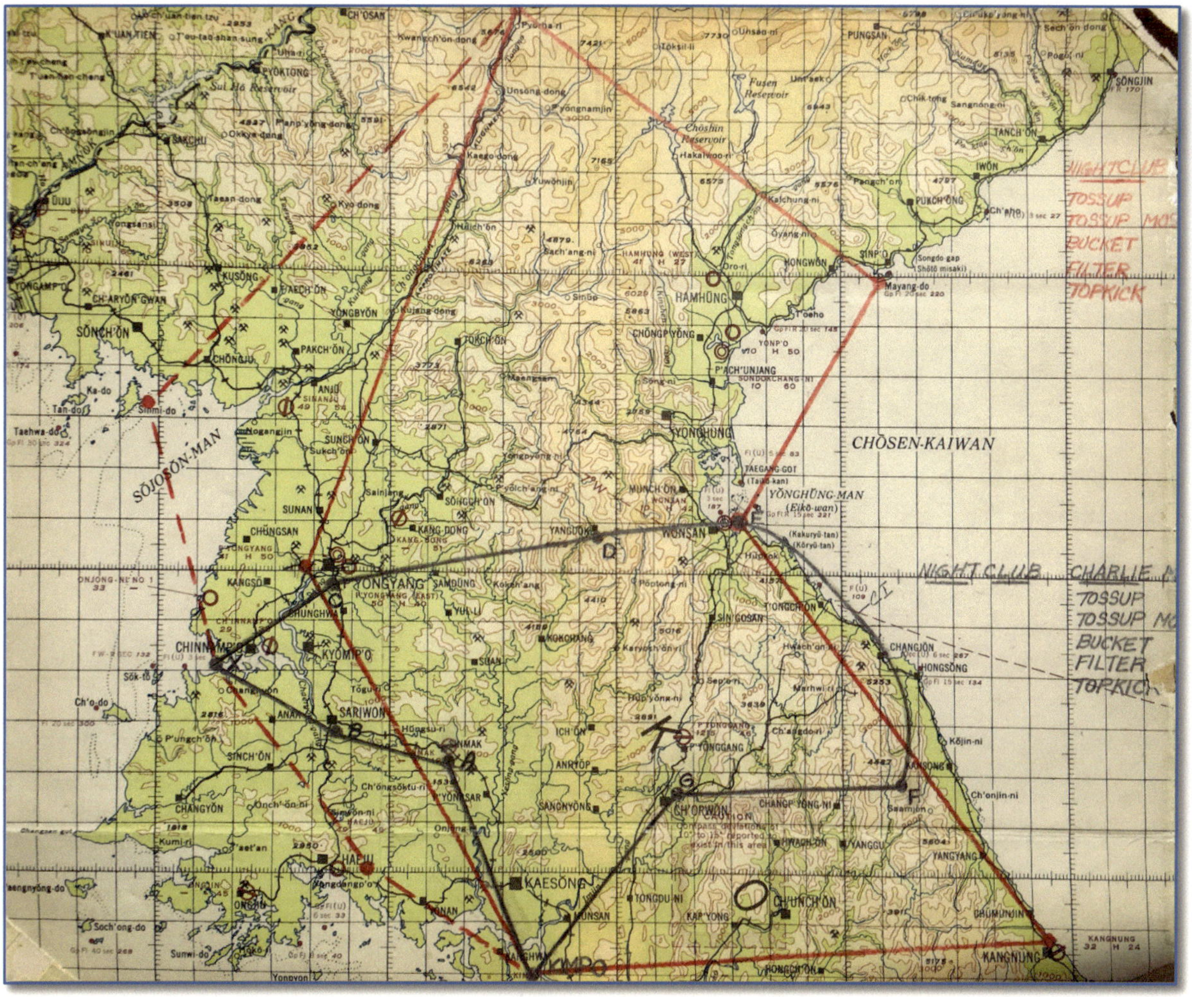

Flight Routes around North Korea to gather "weather" information

Flight routes outside the Korean Peninsula to obtain "weather" information

One memorable flight was on 6 March 1953 during a daylight Charlie mission when I earned the **Distinguished Flying Cross**. The citation reads in part;

> "Despite undergoing intense and accurate enemy radar-controlled anti-aircraft weapons fire over the heavily fortified enemy city of Sinmak, Airman Lehtonen aided in the directing of successful evasive action for the enemy fire and transmitted all weather data obtained along the briefed route to the Joint Operations Control Center. The weather observations provided Fifth Air Force by this mission proved of utmost value to the Fifth Air Force for the planning and execution of effective air-strikes against enemy targets."

In addition to radar-controlled anti-aircraft fire on numerous occasions, we experienced contact with enemy aircraft on two separate occasions. The first was on our 34th mission on 4 May 1953, a night Baker mission as we were proceeding NW to our second checkpoint, I saw a MiG-15 silhouetted against the moon overtake and pass us within 100 feet. The pilot apparently didn't see us. Another time, we were flying at 10,000 feet on a Charlie mission near Yangdok when Kimpo radar reported bogies at our 6 o'clock position and converging. The mountains in this area are about 5,000 feet high and I spotted three MiG's about a mile away. I notified the aircraft commander (AC) that I had a visual on the bogies and he put the aircraft into a steep dive at 425 MPH to get down into the mountains where the enemy planes could not follow. I was sitting on the radio table at the time looking toward the rear and when the AC pushed the nose over, I was still going straight and level and not strapped in but the airplane was headed down. I was thrown around some until I could get strapped in again. We flew down a valley in the mountains with enemy soldiers firing small arms down at us as we flew by. Very quickly we came upon a large rice paddy in an open area with a shack in the middle of it and the weather observer said that he saw someone look out the door and then everyone in the shack came out and jumped into the human excrement fertilized water. When we passed by, I could see six to eight people in the water. They apparently thought we were going to strafe them. Anyway, we made it back to base safely.

Another time, after returning from a stressful mission, I fell out of the aircraft encumbered with heavy combat gear, popped my chute, and fell on my right knee. The AC advised me to go to the hospital to have the knee x-rayed but when you are 19 years old, you shake off the injury saying, "It's nothing." It wasn't until many years later after having experienced knee pain all that time, my knee collapsed one day and I had to have arthroscopic surgery to fix it. Too bad I didn't go to the hospital that day and have the injury reported so my VA medical record would be complete. I finished my combat tour of 50 missions on 19 June 1953, roughly five weeks before the armistice was signed on 27 July 1953 and the 6166 AWRF was deactivated. For having flown a combat tour of 50 missions as an air crew member, I was awarded three air medals in addition to the DFC and other Korean Service medals. On the 50th anniversary of the war, the Korean Government awarded to all combat members who served, the Korean War Service Medal, shown here, presented to my sister Ellen on 7 February 2001 during a ceremony at the Marine Corps Air Facility in Beaufort, SC. I was unable to attend any of the Florida presentations personally because work scheduling conflicts..

Korean War Service Medal

"50 Mission Combat Tour" Complete

My orders to return to the US were dated 2 July 1953 and as combat crew members, we were entitled to fly back to the states. When we got to Brady AFB in Japan, we were informed that due to a scheduling problem we would have to go by ship.

Photo # NH 104018 USS General J. C. Breckenridge during the Korean War

USS JC Breckenridge sailed from Yokohama to San Francisco

The USS J.C. Breckenridge, a WWII Liberty ship shown above, picked us up at Yokohama with the first port of call Okinawa to pick up other returning troops. From there we sailed on to San Francisco but on the way, one of the soldiers fell down a companionway and was killed so we put into Yokahama to drop off the body. The trip took a total of 18 days, our ship being the very first ship to return after the signing of the armistice. However, the City of San Francisco rolled out the red carpet for the next ship that docked because they had no time to prepare for our arrival. I returned home to Lunenburg, Massachusetts for a 30 day leave prior to reporting to my next duty station, Mitchel AFB, New York.

U.S. Air Force Service, Mitchell AFB, NY

I reported to the 1851st AACS Flight, 1802nd AACS Group, Mitchell AFB, NY after Labor Day 1953 as a Flight Check Panel Technician and served in this capacity until the end of my enlistment. After an orientation flight in a B-25, I began crewing the AC-47D, C-45H, and C119 as a panel technician or a radio operator starting on 23 September 1953 in the AC-47D. The mission of the 1802 AACS Group was to conduct flight test activities on air navigation aids in the NE quadrant of the U.S., going as far east as Kindley AFB, Bermuda, north as far as Goose Bay, Labrador, as far west as Duluth AFB, MN, and as far south as Ft. Campbell, KY. Included in this array of navigational aids were VHF Omni Ranges, Instrument Landing Systems, Control Towers, Fan and Bone Marker Beacons and the first Distance Measuring Equipment. VOR testing consisted of flying a 20-mile circle around the VOR facility at a given altitude and recording a visual bearing with a theodolite (similar to a surveyor's transit) and transmitting a signal to the aircraft crossing the radial with a ground-based VHF transmitter, recording on chart recorder in the aircraft the appropriate radial. If the radial fell within acceptable parameters the facility was recertified; if not, the facility was deactivated until maintenance could be performed to correct the problem. I recall a time during the winter standing in a frozen potato field in Ft. Fairfield, ME with the temperature at -30 degrees with a 50 Kt wind performing this VOR check. The road to the facility was plowed but the snowbank on either side of the road was 20 feet high, giving one the feeling that they were traveling in a cold, white tunnel. We were happy to finish that day to spend the night at the Loring AFB transient quarters, which by the way, were excellent because this was a Strategic Air Command (SAC) base and Gen. Curtis LeMay took care of his B-36 airman.

A number of events occurred that I recall with some sadness – others with a sense of accomplishment. In December 1953 I was assigned to a crew to flight check the tower at Kitty Hawk, NC in preparation of the 50th anniversary of the Wright Brothers flight. We completed the flight check and landed at Elizabeth City, NC where we were quartered. The crew was invited to the 50th anniversary banquet and we were distributed among the tables to talk to and listen to the tales of early aviation. I was seated with barnstormers and early air mail pilots who had many tales of flying to remote areas in our country to give airplane rides and of launching off into bad weather to fly the mail. Today I would like to have had a tape recording of those stories and I am sad that I didn't at least get the names of the people who survived what can be best described as the most dangerous period in U.S. aviation history. As mentioned before, our flight check aircraft was an AC-47D "Gooney Bird." This aircraft first flew in 1935 and is very heavy on the controls. Since we had such a large area to service, we all took turns flying from Long Island to the various places to do our flight check work. I learned to fly in that aircraft, not to take off and land of course, but to hold altitude and heading and to follow the air navigational aids to get from one place to another. This informal training led me to solo very quickly during pilot training and to take my solo cross-country flight in a C-150 with 9 hours total time. That accomplishment allowed me to finish up at 35 hours and go for my private pilot flight test with the FAA at Phoenix Sky Harbor airport years later.

During this time, I was fortunate to attend the Flight Check Panel Course given by the Civil Aeronautics Administration (CAA)—currently named the Federal Aviation Administration (FAA)—at Will Rogers Field, OK. I drove from Long Island to Oklahoma City in my 1950 Olds 88 at a time when the highway system was nowhere near what it is today. The two enlisted airmen in the course, Ed Barnett and myself, along with five pilots were trained by the Aeronautical Center staff in the Air Force AC-47D under a cooperative arrangement between the CAA and the Air Force. At the time, the Air Force performed the flight testing of all CAA navigational aids and it wasn't until the late 1950's when the CAA began transitioning the takeover, mostly for high altitude routes and air traffic control facilities. Ed and I rented an apartment for the duration of the course because Tinker AFB had no facilities for us on base. Our landlady was a character. From time to time our duties took us to El Paso or Laredo, TX where we would go across the border to load up on alcoholic beverages to take back to dry Oklahoma. We noticed that our stash seemed to evaporate quite readily so we marked the bottles in pencil and sure enough, we confirmed that she was tapping our supply. She would come in during the day when we were in class and have a nip or two. We never let on that we had discovered her indiscretion but added water to the bottle she was tapping and locked up the rest in our respective autos. We completed our course of study in January 1955 after an intense 12 weeks of classroom work and flying in the AC-47D. Upon my return to Mitchel AFB, I was assigned to a flight check crew to check the VOR at Kindley AFB, Bermuda. In a subsequent trip to Bermuda in September 1955, I recall an incident that today I can look back at and grimace. We had been flying all day and when we landed, found that the mess hall was closed. We did manage to get a cook to feed us and we left to go to the transient barracks. While going to the latrine to clean up for bed, a hand shot out of one of the cubicles holding a bottle of rum. "Here Yank, have a drink." That was my introduction to rum and to the Canadian crew of a 4 engine, De Havilland bomber of WWII vintage. They coaxed me to have another drink and to play poker for much of the night. I had to fly the next day so I broke it off a little early. As soon as I put my head down on the pillow, the whole world started to spin and I got sick to my stomach. I jumped off the top bunk and staggered down to the latrine, hitting roof cross members all the way down the hall. It's a good thing I had to fly the next day because the 100 percent oxygen helped to clear my head so I could do my job. Suffice it to say I didn't have a drink of rum for upwards of 45 years.

I continued to perform flight check duties until my discharge in January 1956, at which point I entered a civilian machinists training program in Ayer, MA and attended engineering school at night. I graduated from Worcester Junior College with an Associate Degree in Engineering in 1960, Fitchburg State College, now Fitchburg State University, with a BSIS in 1966, and Arizona State University in 1972 with an MBA. While living in Arizona during the

1970's, I worked on and received my Private Pilot Certificate (1971), my Commercial Pilot Certificate (1975), my Instrument Rating (1976), and my Certified Flight Instructor Certificate (1979) in Austin TX. As a newly minted CFI, I joined the USAF Auxiliary (CAP) in 1980 and was soon appointed the Squadron Commander of Travis County (TX) Senior Squadron where I learned Emergency Services (Search and Rescue) techniques as well as performing annual flight checks for CAP members in the squadron C-182.

Moving to Florida in 1985, I served as Group Commander and Director of Operations while giving check rides in C-172, C-182, and C-206 aircraft. I was a part time instructor at Patrick AFB Flight Training Center giving instruction in their T-41, PA-28R and PA-28-161 aircraft. I was given the opportunity to serve as Director of Operations for SE Region CAP where I was able to travel to adjacent states in this role.

As SE Region DO I was able to assist with flight check schools, one such trip to Dobbins AFB, GA. During this trip I was able to fly one of the few remaining GCS instrument approaches. What a kick! The photo below shows me with the FL Wing C-206 during a training mission in Ocala, FL.

Lieutenant Colonel Lehtonen with C-206

"Joyce and I reside in The Villages, Florida's Friendliest Home Town."

"Sing and make music from your heart to the Lord, always giving thanks to God the Father for everything, in the name of our Lord Jesus Christ." (Ephesians 5:19-20 NIV)

Intelligence Gathering

The Cold War saw the invention and promulgation of a variety of intelligence gathering systems and capabilities.

Lockheed U-2

The Lockheed U-2, nicknamed "Dragon Lady," is an American single-jet engine, ultra-high-altitude reconnaissance aircraft operated by the United States Air Force (USAF) and previously flown by the Central Intelligence Agency (CIA). It provides day and night, high-altitude (70,000 feet; 21,336 m), all-weather intelligence gathering. The U-2 has also been used for electronic sensor research, satellite calibration, and communications purposes.

SECTION 3: THE COLD WAR

The COLD War was a state of geopolitical tension after World War II between powers in the Eastern Bloc (the Soviet Union and its satellite states) and powers in the Western Bloc (the United States, its NATO allies and others). Historians do not fully agree on the dates, but a common timeframe is the period between 1947, the year the Truman Doctrine (a U.S. foreign policy pledging to aid nations threatened by Soviet expansionism) was announced, and 1991, the year the Soviet Union collapsed.

The Cold war produced competitions like the Space race weapons and target acquisition, arms race.

In the period from the end of the Korean war to the collapse of the Soviet Union, spy technology produced a series of innovations including the ultimate spy in the sky satellites able to read a license plate on a car from 160 miles up in space, and we were able to put nuclear weapons into orbit with the ability to drop them anywhere.

Proliferation of Nuclear weapons and multiple nuclear weapons tests had kids in school practicing nuclear war preparation by *hiding* under their desks not looking out the windows; at least one author was able to stand on the ground outside the school and feel the Earth shake when an atomic bomb was tested in Nevada 500 miles away. We had aircraft on alert and B-52 bombers in the skies 24/7 prepared to rain down nuclear destruction on a moment's notice.

Army soldiers and their families stationed in Europe were on alert with Strategic readiness kits STRAC or STRIKE kits with blankets, flashlights and fuel for their cars in case the *balloon went up*.

Source: Various

3

Eugene (Barney) Barnhart
Major
United States Air Force
Cold War and Vietnam War

A young Captain Barnhart

I volunteered to go to Vietnam in October 1968. I had my choice of assignments and I chose the RF-4C Phantom. This had been my first choice from pilot training. Obviously, I did not receive my first choice from pilot training since I was assigned to the B-52s at Carswell Air

Force Base in Fort Worth Texas. After flying the B-52s as a copilot for a couple of years, the final year as the copilot for the chief of standardization and evaluation, I was finally looking forward to fighter planes.

Flying B-52's instead of Fighters for 2 years

I would be flying RF-4C phantoms. Although it was not a true fighter since the mission was tactical reconnaissance, I did not have to drop ordinance of any kind. I watched the Robert Stack movie "Alone, Unarmed and Unafraid" and soon found out that it was a very hazardous mission of gathering intelligence over enemy territory without the benefit of any protective weaponry on board. I was eventually assigned to the 16th tactical reconnaissance squadron at Tan Son Nhut Airbase in Saigon.

McDonnell Douglas RF-4C Phantom II

The fighter pilots thought we were nuts as I was informed many times at the bar. We relied on speed and maneuverability to avoid enemy fire when gathering our intelligence and filming the bad guys. The RF-4 was loaded with cameras and various other sensors such as Infrared (IR) and side-looking radar to detect enemy positions and hiding places. Initially I spent many flying hours doing "point and click" missions, photo strips and night IR area covers. This intelligence information or 'intel' was then downloaded upon my return and examined by photo interpreters.

The intel was then passed on, with photos and reports, to the seventh Air Force headquarters staff so they could send troops or bombers to destroy the bad guys or try to determine their next tactical moves.

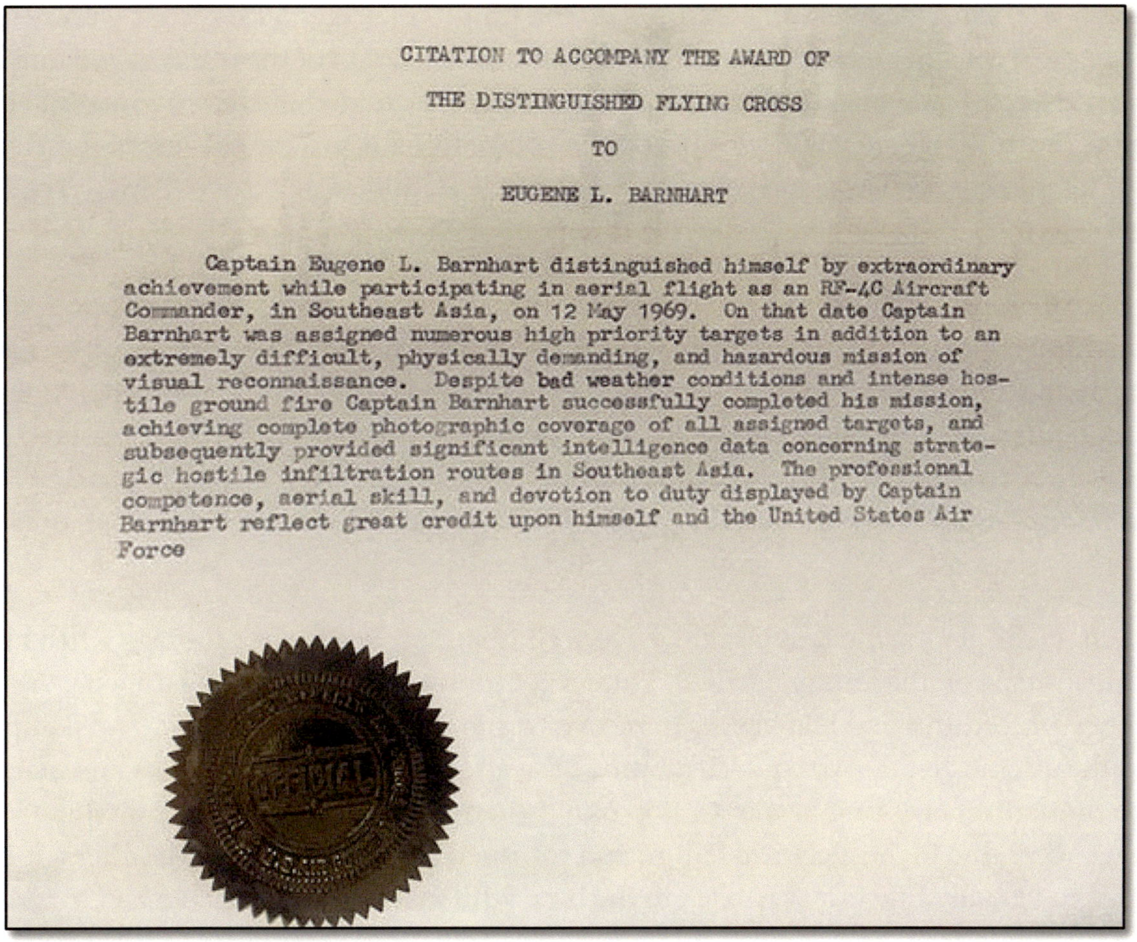

CITATION TO ACCOMPANY THE AWARD OF

THE DISTINGUISHED FLYING CROSS

TO

EUGENE L. BARNHART

Captain Eugene L. Barnhart distinguished himself by extraordinary achievement while participating in aerial flight as an RF-4C Aircraft Commander, in Southeast Asia, on 12 May 1969. On that date Captain Barnhart was assigned numerous high priority targets in addition to an extremely difficult, physically demanding, and hazardous mission of visual reconnaissance. Despite bad weather conditions and intense hostile ground fire Captain Barnhart successfully completed his mission, achieving complete photographic coverage of all assigned targets, and subsequently provided significant intelligence data concerning strategic hostile infiltration routes in Southeast Asia. The professional competence, aerial skill, and devotion to duty displayed by Captain Barnhart reflect great credit upon himself and the United States Air Force

One day several months into my tour, I was approached by my squadron commander, George Sauls, with an opportunity to volunteer for a new classified mission called Visual Reconnaissance (VR). The seventh Air Force was having difficulty keeping up with the ever-changing intel along the Ho Chi Minh Trail coming out of North Vietnam. They were going to assign six aircrews from the 12th and 16th tactical reconnaissance squadrons to fly it daily. This was to be the only missions they flew! It was a special mission that involved being on station in southern Laos early in the morning and then flying high-speed, low level along the Ho Chi Minh Trail until evening. The final sortie of the day would land around dark at Tan Son Nhut airbase South Vietnam. The normal reconnaissance sortie time was about two hours. The VR sortie was usually closer to 3 to 4 hours with at least one air to air refueling over Laos with another pass down the trail on the way home. We would fly north out of Tan Son Nhut to around Attapeu, Laos near the Cambodian border and then descended to our VR altitude and speed along the trail. We would fly about 480 to 500 kts in a constantly jinking maneuver at low altitude as we surveyed the trail. The enemy would normally not hear us coming at that speed until we had already passed them. That did not keep him from calling ahead to his buddies to let them know we were en route on the trail so they could fill the air with weapons fire in hopes of catching us as we flew over. Because of the stress of the mission, we limited each run on a road to about 10 minutes. Consequently, we divided the entire route into three segments. Entry was the Air America portion in southern Laos to a point West of the A Shau Valley. The mid segment was to Tchepone, West of the Demilitarized Zone (DMZ). The final segment was into the passes and valleys leading out of North Vietnam.

Shortly after I started flying VR missions, I received an assignment to pick up a small nighttime IR cover strip on my way home from Laos. It would normally have been an easy mission after a tiring couple of hours jinking, "yanking and banking," on the trail. ("Jinking" or "yanking and banking" was 'pilot-talk' for a method of constantly changing the heading and altitude of the aircraft with short partial turns left, right, up, and down. This made them a more difficult target for the enemy. Remember, the RF-4 had no guns or armament, only cameras and sensors.)

After completing the first target we were climbing up to a higher altitude when all the electrical systems on the airplane failed! Things get pretty black when this happens! We had emergency procedures that allowed us to recover some of the electrical power and equipment to keep the aircraft flying. We contacted home base after some radio difficulties and alerted them to prepare an arresting barrier to stop our airplane. We had no means to let them know where we were due to the electrical failure and fortunately even though it was after sunset, we were able to visually find our way back to the base without radar contact. We had to use the emergency methods to get our gear and flaps down. To further complicate our problems the nose gear would not extend nor would the flaps go down. We tried to jar them down by shaking the airplane and jerking the stick back but to no avail. We found the field and had been circling during the process of trying to get the nose gear and flaps down. We then got a low fuel warning that indicated at most five minutes before we flamed out and would have to eject.

While all this was happening, they were going to foam the runway because of having to land with the nose wheel still in the in the wheel well. When they heard about our fuel situation, they informed us that we would have to land immediately. All of this so far adds up to five different things each of which would have constituted an emergency. When I was told that I would have to put it down immediately, I was about 45 seconds from the runway at 160 MPH. They had set up arresting cable on the selected runway for my landing before I got the unsafe nose gear warning. An arresting cable is a single steel cable that can catch the tail hook. But with no nose gear one cannot land to an arresting cable because the cable will travel up over the nose of the aircraft and probably cut off the pilot's (that would be me) head. I had to swerve several hundred feet over to the parallel runway that had no cable to avoid yet another catastrophe. Of course, the other runway was not foamed. I made a beautiful high-speed landing (even if I do say so myself) and gradually lowered the nose considerably lower than natural to the runway. That is when the sparks started the fly and my centerline external fuel tank blew up. To make matters even more interesting, I did not have steering or normal brakes. I had to try and steer by using my emergency brakes. We touched down at about 160 mph and started sliding to the left off the runway at about the 4000 foot marker. Luckily, we missed several ditches and the remote-control shack between the runways. All this time we were being trailed by fire caused by the sparks and residual fuel.

When we finally came to rest adjacent to the 5000 foot marker, we went through our emergency shutdown procedures and evacuated the aircraft in record time. When I hit the ground, my legs were going full speed. I came to a sudden halt about 15 feet from the aircraft when my nylon lanyard attached to the ejection seat stopped me in mid-stride. I managed to free myself with the help of my backseat Reconnaissance Systems Officer (RSO). All I could think of was the flames beginning to get awfully close to those other gas tanks. I was afraid the whole plane would explode before I could get far enough away. The fire truck and rescue helicopter were there almost immediately and got us out of harm's way. One of our people in the control tower said the first indication he had that we had landed with no lights was when he saw the flames following the aircraft down the runway. Without lights on the aircraft they were never sure where we were until they saw flames on the runway. Our squadron commander said he wouldn't have given us a dime for our chances of surviving when he saw us land and slide along. It turned out after the investigation to be all related to multiple mechanical failures that caused the problems. All of our film was recovered and used to provide important intelligence for the following day's missions. This was just one of many demanding missions that resulted in my receiving nine Air Medals and the Distinguished Flying Cross for extraordinary achievement during my tour in Vietnam.

...The aircraft was salvaged and could be flown again.

The fire extinguishing foam from the fire rescue truck prevented the aircraft from burning to the ground. The aircraft was salvaged and could be flown again.

The long photo below is a photo-mosaic that shows the track of the aircraft as it slid off the runway, into the median, missed a radar control box, and skidded to a stop near the 5000 foot marker- -The dark burn spot, left in photo, is where the aircraft came to rest. Our RF-4C slid nearly a mile on the centerline fuel tank and came to rest at the left in this photo showing the dark triangular shape of the burn spot in the median between runways.

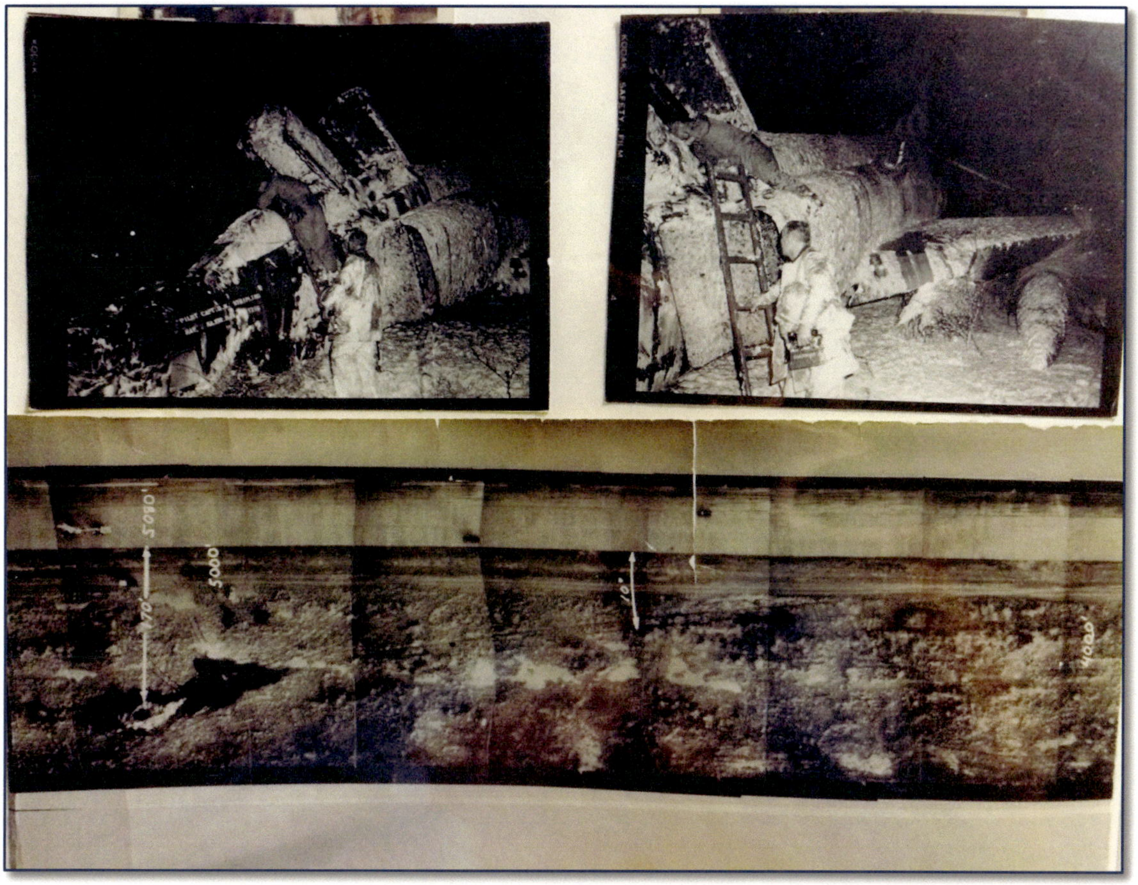

Photo-mosaic montage demonstrating slide path off runway

Lower photo shows path of the aircraft as it slid off the runway and stopped at the black patch (lower left).

I went out to see the wreck the next day and had my picture taken beside it just so people would believe what had happened, and that I was there.

Captain Barnhart standing next to wreckage

The next day the aircraft was lifted out and taken to a maintenance hangar for repair.

Airlifting the RF-4C wreckage for repairs

Here I am on the side of Interstate 4 in Florida near the Orlando airport with a memorialized MacDonald- Douglas F-4 like the one that managed to save our lives some 50 years ago.

Barney Barnhart today

Barney poses with a B-52 that he flew at Carswell AFB in 1966, it's now on display at Orlando International Airport

4

Dennis P. Derrig
Lieutenant Colonel
Air National Guard (Ret)

Cold War & Vietnam War

Lieutenant Colonel Dennis Derrig

Dennis P. Derrig flew fighter jets during and after the Vietnam era. His life as a combat fighter pilot began immediately following the Gulf of Tonkin incident.

Dennis enlisted in the Air National Guard in July 1958. After completing two years of his college education, he was accepted into U.S. Navy flight training in August 1961. He completed his flight training in March 1963, went to A4 training with the Replacement Air Group in California, then joined VA-55 (an Attack Squadron of the U.S. Navy) flying the A4E on combat missions into North and South Vietnam. He got out of the Navy in 1966 to complete his college degree.

Douglas A-4E Skyhawk

Dennis P. Derrig was awarded the Distinguished Flying Cross in 1966.

Citation:

"The President of the United States takes great pleasure in presenting the Distinguished Flying Cross to Lieutenant Dennis P. Derrig for heroism and extraordinary achievement while participating in aerial flight as a Pilot flying light jet attack aircraft in Attack Squadron Fifty-five, a unit of Attack Carrier Air Wing Fourteen embarked in USS Ranger (CVA-61) as a part of Seventh Fleet on 31 January 1966. As Wingman in a flight of four A4E Skyhawks on an armed reconnaissance mission over North Vietnam, Lieutenant Derrig completed the mission in spite of low ceilings and limited visibility. He successfully and accurately conducted low altitude bombing attacks on the ferry landing south of Dong Hoi inflicting heavy damage and starting several secondary fires. Intense anti-aircraft artillery in the area resulted in major damage to his aircraft and loss of his Section Leader's aircraft. Disregarding his own safety, Lieutenant Derrig conducted searching passes in the target area until the downed Pilot was located and acted as RESCAP until the Pilot was rescued. Lieutenant Derrig's aggressiveness, courage, and selfless action were in keeping with the highest traditions of the United States Naval Service."

Not stated in the citation is the extreme flying skill Dennis demonstrated while waiting for his wingman to be rescued: he stayed on station until his fuel was nearly exhausted, climbed up to a refueling aircraft, refueled and went back to keep an eye on his wingman and coordinate the rescue. He had to refuel again once the rescue was made in order to have sufficient fuel to make a landing on his home aircraft carrier.

In 1968, he rejoined the North Dakota Air National Guard as a pilot and flew with them for the next 22 years in a distinguished and unique dual career as a pilot in several different kinds of aircraft of that era, the C-54, T-33, F-101, F-102, and F-4D, and working as a Civil Engineer.

Douglas C-54 Skymaster

When he finished his military career, he had over 1000 hours in the A-4E, 1000+ hours in F-101, and another 1000+ hours in the F-4D. In addition to his Distinguished Flying Cross, he has 15 awards of the Air Medal.

Dennis Derrig at DFC Society Chapter meeting

5

Frederick K. Koerner
Lieutenant Colonel
United States Air Force (Ret)

Cold War & Vietnam

A young 1st Lieutenant Frederick Koerner

As a young boy I had a real interest in airplanes, as had a lot of kids my age. I would build models out of balsa wood covered with a tissue paper and power them with a rubber band. I would fly that poor thing until it only had one more crash-landing left.

Then I would tie a M80 firecracker on it, light the fuse and send it off one more time, never to return.

As I grew older, I became interested in things electrical and mechanical so I selected courses in high school that would prepare me for college to study engineering. I applied and was accepted at West Virginia University and started my freshman year in the fall of 1959. I entered the school of Electrical Engineering and selected the Air Force Reserve Officers Training Corps (ROTC) program (ROTC was required). However, after 4 semesters I realized that I was not ready for college. I just did not have the discipline and ability to establish the kind of study habits required for my chosen field of study.

I dropped out of college and spoke with the local Air Force recruiter about what the Air Force might have to offer me since I had two years of ROTC. He told me they had a school called the Aviation Cadets that trained either pilots or navigators, depending on the forecasted needs of the Air Force. He told me that if I wanted a shot at the program, I would have a much better chance of being accepted from "inside" the Air Force training environment. So, in late 1961 I enlisted in the Air Force with the understanding that my request for this school would be given serious consideration but no guarantees. I went through basic training at Lackland AFB in Texas and was put on "Aviation Cadet hold" and waited to see if they would accept me into the program which they did about 3 weeks later. I was put on a bus for James Connelly AFB in Waco, Texas and was on my way. They were only training navigators then as the pilot training program had been terminated. But I didn't care, I was going to fly!

After receiving my commission and navigator wings from the Aviation Cadet program in February 1964 I was assigned to the 960th Airborne Early Warning and Control Squadron at Otis AFB in Falmouth, Mass. I spent 2 years there in the EC-121 flying racetrack patterns out over the Atlantic looking for the Russians and when it came time to be reassigned, I volunteered to go to Vietnam.

I was given orders to go to Alexandria LA for a special type of ground training which turned out to be learning to pinpoint radio transmitters on the ground with specialized equipment on the EC-47. This was the basic cargo model with 2-3 radio operator consoles and a navigator position installed. After this ground training we were sent to Hurlburt Field in Florida to learn the equipment and to become proficient in it's use. The waiting, traveling, and training for this mission took about six months before we were finally assigned to the plane in Manchester, NH that we were to ferry over to Saigon, Republic of Vietnam (RVN). The trip went from Manchester to Adak, Alaska where we waited several days for a favorable wind factor to get us, and our ferry tanks, to Midway Island. Then on to Wake, Guam, Clark and Saigon which took 21 days total. Our plane had the 'latest' in navigation aids including a hand held sextant, a drift meter and a Doppler computer which went 'belly up' after about five hours into the leg from Adak to Midway. This left us with just the sextant and drift meter to figure out where we were, and plotting our fuel remaining over distance to go graph with a razor-sharp pencil.

I spent about two months at Tan Son Nhut before getting transferred to Nha Trang where I made Captain. We were the cadre in this mission and new crews were being sent out from Tan Son Nhut to our other bases as we staffed up. A typical 7-hour mission included a pilot, co-pilot, flight engineer, navigator and 2-3 radio operators. After arriving at our assigned area, we would get a Doppler computer set from which to plot our current position. We would continue to update/correct the Doppler computer (remember the 'belly up' equipment) from time to time. As we patrolled our area at an altitude of five to seven thousand feet, the radio specialists would be listening in on radio traffic and would call out 'lock on alpha, or bravo, or charlie, etc' when they found a person/transmitter they were interested in, and I would start plotting the location of that transmitter. If it was someone they were really interested in they would say that the target was a priority one (or something to that effect). Someone of the highest priority was classified as a 'priority one Alpha' which were usually few and far between. I vaguely remember a few such targets, one of which had to have been the one on 2 September 1967 in the Bong Song area for which I was awarded the Distinguished Flying Cross.

Frederick Koerner and his flight crew

I never knew where the information I gathered from each mission ended up and never received any feedback on the intelligence we were able to gather. All that went with/to the radio operators and their section. I do remember the Bong Song area as it was not that far up the coast from Nha Trang, both of which were along the coast but I don't remember all the details of this particular mission since, to me, they were all the same with equal effort given to locate all transmitters. As I said, we flight crew members were never given after action reports on the intelligence we gathered so I didn't have a clue as to the results of my locating that particular transmitter. After my tour was up, I rotated back to the 'States a week after the Tet Offensive started and entered the inactive reserves. And sometime after my leaving active duty I received the notice of the award. I later joined the 167th TAS, an Air National Guard C-130 unit in Martinsburg, WV and spent 16 years with them, including flying Desert Shield missions before retiring.

I received notice of being awarded the DFC for the Bong Song mission. The citation reads as follows:

> "Captain Fredrick K. Koerner distinguished himself by extraordinary achievement while participating in aerial flight as an EC-47 Navigator near Bong Song, Republic of Vietnam on 2 September 1967. On that date, Captain Koerner flew a highly classified, long and arduous reconnaissance mission over hostile terrain and under adverse weather conditions. Through the selfless application of his outstanding professional skill, ingenuity and perseverance, Captain Koerner's efforts resulted in the collection of the highest priority reconnaissance data which aided immeasurably in the fulfillment of complex and urgent intelligence requirements in Southeast Asia. The professional competence, aerial skill and devotion to duty displayed by Captain Koerner reflect great credit upon himself and the United States Air Force".

After several years in the plumbing and pipe fitting trades, in 1980 I went back and finished my college degree in Business Administration and Data Processing. I worked with computer systems until I retired in 2005 while working at Florida State University in Tallahassee, FL. My wife and I retired to Goodyear, AZ for 7 years but then moved to The Villages, Florida in 2012. The activities I have enjoyed outside of my work and profession have been motorcycles, scuba diving, and flying several different types of single engine aircraft. I now enjoy boating, golf and tennis.

Fred Koerner today, flight suit still fits!

6

Dwight Alan Moore
Lieutenant Colonel
United States Air Force (Ret)

Cold War & Vietnam

Lieutenant Colonel Dwight Moore

When I was a kid growing up in Kansas, I'd ride my bicycle on my paper route and dream of flying. Not having much money, I lived at home while going to Junior College and then got student loans to attend Kansas State College of Pittsburg, KS. There I met the love of my life, Nona. After graduating in 1968, we spent the summer in Cripple Creek, Colorado working at a hotel until my Draft Board in Kansas City asked me to report for a physical. That was when I enlisted in the Air Force and signed up for pilot training.

I left for Officer Training School in San Antonio, TX in September 1968, followed by pilot training at Reese AFB in Lubbock. In December, Nona and I were married. In flight school in Lubbock, we trained on the Cessna T-41 (Mescalero) a propeller-driven training aircraft, then the T-37 (Tweet) a training jet, and finally the supersonic T-38 (Talon). Each step was to determine if you could handle a more capable airplane. At the end of flight school, we were given a list of airplanes available and we got to choose based on our class rank. My dream was to fly the F-4 (Phantom) Fighter Jet, but there were none available. So, when they gave me the silver wings I had coveted since I was a kid, I chose to go to Carswell AFB in Ft Worth, Texas, the Strategic Air Command (SAC), and began flying the B-52D (Stratofortress) long-range, subsonic, strategic heavy bomber. Once checked out on the aircraft, I was placed on a crew, sat nuclear alert, and then left for an "Arclight" tour (3 months of flying combat).

In Vietnam, the B-52s flew missions out of Anderson AFB on Guam, which was an 11-hour round-trip flight, or out of the Royal Thai Air Force base at Utapao (UT), a huge base in Thailand with the only runways long and wide enough to handle B-52s. UT was only a four-hour combat flight. Our normal schedule at UT was to fly three days with the same take off time, three days with a one-hour later take-off time, and then have a day off or be on standby, gradually working our way around the clock. Two hours ahead of flight time we would pick up our gear, go to our briefings and get the tail number of the plane we would fly. Once at the airplane with systems checked out, we would start the engines and see if it was going to fly. If there was a problem with the plane or one of the engines, we would shift to the back-up aircraft. There was always one waiting. The eight-engine B-52D fully loaded weighed up to 450,000 pounds, half of which was fuel.

During my first tour, I flew 50 combat missions as the co-pilot. We mostly bombed in South Vietnam and, later, along the Ho Chi Minh Trail network in Laos and Cambodia. That was a low-threat area, where we did not face the kind of heavy AAA (Anti-Aircraft Artillery) also called Triple-A and SAMs (Surface-To-Air Missile) we later did over North Vietnam. UT had only D-Model B-52s based there. Guam had some D models plus the newer G models, which carried fewer bombs and did not have the latest electronic counter-measure gear. When we flew out of Guam, it was the same two-hour briefing and debriefing, sandwiched around eleven hours in the air. Guam was 2,000 miles from anywhere, including the targets, so we had to do inflight refueling over the Philippines.

During my three subsequent 6-month tours over Vietnam I flew 166 missions as the pilot and aircraft commander. As Aircraft Commander, I was the leader of a 6-man crew. My squadron commander saw me as the one person he held responsible for the aircraft and mission accomplishment. My crew saw me as their boss and team leader, responsible for training them to work together effectively and, if possible, getting them promoted and helping advance their careers. The pilot had the final word on everything while in the air but had to trust his team. The wives of the crew were more critical: Was I a good pilot? Was I precise on instrument flying in the weather, and when we had to do inflight refueling, did I always get the gas? Did I know the aircraft and its hundreds of systems so thoroughly I could resolve or work around system failures or malfunctions or minor emergencies in the air to get the job done? Did I fly well under stress and, most importantly, could I be trusted to always land it smoothly and bring their husbands back safely every time?

I would note that in Vietnam we primarily carried the Mark 82, an unguided, low drag, general-purpose bomb, part of the United States Mark 80 series) known as a dumb bomb. Developed in the 1950s, it weighed 500-pounds, give or take, and cost around $2,000. With mechanical nose and tail fuses, and 197-pounds of explosives inside, it made a hole in the ground roughly 40 foot in diameter and 30-foot deep and had a fragmentation radius of 100 yards (94 meters) and a blast radius of 14 yards (where air pressure from the blast can cause injury). There were no laser or GPS precision guidance systems back then. We carried dumb, gravity bombs. Lots of them. We carried 84 of the Mark 82s in the bomb bay and usually 24 of the larger 750-pound Mark 117s (air dropped demolition bombs) outside on the wing racks, for a total of 108 bombs per trip.

Our "precision guidance system" was the skill of our 6-man crew to identify an invisible point 30,000' to 35,000' in the air above the target and release the bombs so they all fell (roughly 2 minutes to impact after release) in the "target box." My job was to get the aircraft to that invisible point. At 600 knots per hour, if we were even a half degree off the correct heading, we'd put 108 very large craters in a strip a quarter of a mile wide and up to 8 miles long in the wrong place. However, since most of my 216 combat missions were over dense jungle along the Ho Chi Minh Trail which the North Vietnamese used to transport supplies into South Vietnam, dumb bombs were fine and pinpoint accuracy was not critical.

All that changed in December 1972. We still had only dumb bombs, but the North Vietnamese had withdrawn from the Paris Peace talks. Congress had promised to defund the war when it returned in January 1973 and the United States did not have an agreement guaranteeing the return of our Prisoners of War. President Nixon presented North Vietnam with an ultimatum - return to the talks or there would be consequences. They did not return. As a result, on 18 December, the President ordered the United States military to begin a "Maximum Effort" bombing campaign against North Vietnam. The effort lasted 11 days and became known as the Linebacker II campaign (LB-II) - America's last great air battle of the Vietnam war.

In mid-December 1972, it became obvious around our base at U-Tapao, Thailand (UT), that something was up. Our normal round-the-clock aircraft takeoffs and landings stopped, and maintenance personnel worked non-stop getting bombers loaded and ready to fly. For us, "Maximum Effort" for the first night translated into 129 B-52 aircraft, grouped in three Waves, plus 102 Air Force, Navy, and Marine support aircraft (fighters, tankers, air rescue, etc.) attacking North Vietnam. B-52s normally flew in a triangular, three-aircraft formation called a "Cell" and there were three cells per wave. (Each Cell used a different color for its call sign). On 18 December 1972 my crew flew as "Orange 3" (3rd aircraft) in the Cell in the 3rd wave. This would be the first of our six night-bombing missions during LB-II.

At the Preflight Briefing, you could have heard a pin drop in that room when they announced our target was "Radio Hanoi" - followed by a lot of "Oh, shit!"s. It was not your local FM or AM station. It was the control site for guiding North Vietnamese interceptors, MiG-21 (Mikoyan -Gurevich- or Fishbed) supersonic fighter aircraft, against their favorite targets: B-52 bombers. It was located on the outskirts of Hanoi and defended by over 1000 surface to air missiles and hundreds of Anti-Aircraft-Artillery (AAA) guns. The map on the wall in the briefing room showed roughly fifty dots, which were the ***known*** SAM sites along our inbound flight path. Then, after we dropped our bombs, we would come back out over them again.

The SAMs supplied to North Vietnam by Russia were the SA-2 (Guideline- or S-75 Dvina), a Mach-3.5 rocket with a 288-pound blast-fragmentation warhead guided by ground-based radar. Think of a telephone pole flying up at you at three times the speed of sound. Because of the low cloud cover we experienced during the first night of LB-II, when we looked ahead of us toward the ground, a SAM launch first appeared as a circular white glow in the clouds, followed by a bright dot of light growing in the center as the rocket shot upward. If the bright light did not shift across the windscreen and only grew bigger, it was headed for you.

This was our first "PRESS-ON" mission. That meant if the bombs would be dropped on the target, even if the aircraft had no defenses or would likely be lost in the effort, the crew could not abort or turn back. If they did, they were told when they returned to the base, they would face court-martial charges and possible imprisonment. All LB-II missions were "PRESS-ON". The intelligence briefer said we would be in range of MiG aircraft for 632 miles, roughly 80 minutes, and within range of known SAM sites for 272 miles, about 35 minutes. The only AAA capable of reaching our altitude were 100mm guns but their cone of fire was small, thus the high-altitude bombing. Crews had no way of knowing how much AAA was fired at them. Fortunately, AAA was not a significant threat to us although it was for our brothers in the support aircraft.

The most important part of our Preflight Briefing was the target studies given to our Radar Navigator and Navigator. There were 14 known POW (Prisoner of War) camps with our guys in them in North Vietnam. Accidentally hitting one of them because of crew error could not happen - period! The Radar Navigator had to find the correct aiming points on his radar scope to enable the airplane's computers to direct us to that exact point in space where the release of the bombs would result in all bombs hitting only the target.

To ensure an accurate bomb release, the aircraft was required to fly straight and level at the correct altitude, airspeed and heading for one full minute before release, no matter what was being fired at it. That straight run, with the bomb bay doors wide-open, followed by the subsequent sharp turn to get out of there once the bombs were released was when we were the most vulnerable. That was where we lost the largest number of B-52s.

One of the reasons for those losses was a major flaw in our tactics, which Headquarters SAC refused to change. All flights flew to the target spaced out along the same path at the same altitudes, same airspeed and then made the same sharp banked turn as soon as they dropped their bombs past the target. **We had become predictable**. The North Vietnamese (NVA) were no fools, and that became their aim point. Our Electronic Counter Measures (ECM) were very good. The radar jamming energy put out by a Cell of three B-52s blinded their radars, leaving their screens with nothing but a fuzzy white cloud. However, the ECM antennas were aimed at the ground. When we made our sharp, banked turns, that gear was now aimed more sideways, and the aircraft was partially exposed, allowing a radar return to the SAM site. When a number of other planes passed over the same spot and made the same turn, the NVA marked the location. They knew the time between Cells as they passed over the same target, so once the lead Cell passed over, they would fire a barrage of SA-2s (Surface to Air Missile called Guideline, known for shooting down Gary Powers in a U-2 Spy plane in1962) at that spot. It was effective and brought down Rose 1 (lead aircraft) in the Cell behind us. We complained, but SAC wouldn't change the plan.

We survived our first mission. By our count, we avoided 13 SAMs fired at our Cell as we flew in and came back out. It was a busy time in the cockpit. Once we were outbound from the target "Red Crown," the U.S. Navy ship off the coast monitoring the attacks, told us that an enemy MiG-21 was attacking bombers ahead of us and was headed our way. Thanks to the fighters accompanying us, by the time he flew by us he had four US F-4 (Phantoms) chasing him. We lost three B-52 aircraft that night. We were now in a serious showdown with the North Vietnamese.

Missions two through four were less exciting for us, but not for the North Vietnamese. The Thai Nguyen Thermal power plant, the Bac Giang Transshipment Point, and the Lang Dang Railroad repair yards disappeared under a hail of dumb bombs. No SAMs were fired at us during those missions, so life was good. However, while no aircraft were lost the second night, six B-52s were lost the third night and a total of 15 during LB-II, plus 6 support aircraft. Ten of the B-52s went down over North Vietnam. Of the 61 crew members aboard, 28 died and 33 became POWs.

Unlike those wonderful photographs from World War II, we did not fly the same airplanes every day. The mechanics stayed with the same planes, but the pilots did not. They put us in whatever was available, sometimes filling in with other crews based on wherever you were needed. For my fifth mission my gunner and I went up with another crew, and the target was the Van Dien supply storage depot and vehicle storage area 5 miles south of Hanoi. The aircraft had a few minor problems with its navigational equipment and air conditioner, which ran full hot during the mission. We were the third ship in Silver Cell, and we were going in 9 minutes behind Ash Cell. My regular radar navigator, Bill North, and my Electronic Warfare Officer, Dennis Anderson, were substituting on Ash 2. They were hit by a SAM as they began their post-target turn, and while I was flying inbound, I could listen to their efforts to keep their ailing B-52 in the air long enough to reach Thailand and bail out. Thankfully, all the crew made it out safely and without serious injury.

Although we were only within the range of SAM batteries for 120 miles, or 15 minutes, I heard the calls on the radio about SAM launches as soon as we turned toward the target. It was now our turn. At 60 seconds out, our orders were to fly the aircraft straight and level to allow the bombing computers to stabilize and guide the bomber to the release point. That meant no evasive maneuvers.

It was on that final leg that I saw three SAMs pierce the low cloud cover on my side, heading upwards. The co-pilot saw three SAMs launched on his side, gunner called two more were airborne aft of the aircraft, and three new ones appeared off the nose. On the radio, Red Crown broadcasted "SAM Launch…SAM Launch, Bullseye (which was Hanoi)," but I barely heard him. Sixty seconds to go. Between the radio calls was my two-word prayer, "Oh Lord." Eternity. Thirty seconds! The target was only a few miles from the Hanoi Hilton, the Zoo, and the Skid Row POW camps, so there could be no mistakes. Finally, "5, 4, 3, 2, 1, Bombs away," and, "Turn, Pilot!"

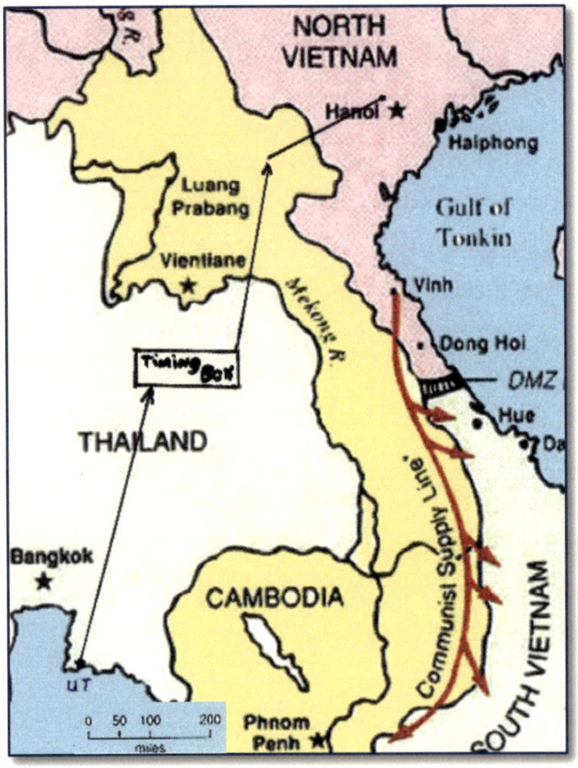

My aircraft responded beautifully as I rolled into a 45-degree banked turn, pulled hard on the yoke to keep it level, and brought the nose around. Outbound at last! By our count, eighteen SAMs had been launched at our Cell, and the only safe path through them appeared to have been the straight-and-level bomb run. We were outbound when a MiG-21 tried to attack the Ivory and Opal Cells, but we were clear. Back to Utapao and a beer, thank God. My 5th mission down.

Our final mission of LB-II was on 29 December 1972. We were flying as Maple 3 against the Phuc Yen SAM support facility and storage area, ten miles northwest of Hanoi. It was our performance on this flight for which I was awarded the Distinguished Flying Cross.

The mission was one of the most complex of LB-II with aircraft approaching the target from different angles, leaving by different routes and altitudes. Three Waves of three Cells (27 airplanes) each attacked Phuc Yen. The release time of each Cell exactly matched those of its counterpart Cells in the other two Waves. The post target routing had two of the Waves crossing tracks, separated only by altitude. A delayed release by a high sortie would result in the bomb train falling through the path of the other Wave. This was before the invention of collision avoidance systems so it required closer timing than on any previous day. Additionally, Maple Cell, flying in from the northwest, had to go out on a route it held in common with Walnut Cell, which had identical target times. As a result, maintaining the correct assigned altitudes, headings and airspeed were critical. To complicate matters, Wine Cell had only two aircraft and merged with Walnut for an unusual five-ship Cell. Bomber Waves were preceded in the target area by F-4s dropping chaff (unfortunately it didn't help much because the chaff corridor drifted with the wind). There were several sets of F-4 fighters doing MIGCAP (MiG Combat Air Patrol) escorting the Cells in case of MiGs.

Between six and eight SAMs were launched at our Cell during the bomb run. At the post-target turn, I was to descend 1,000 feet while in the turn. Great Idea. Half way through the turn, the cockpit, which was normally pitch black at night, became bright enough to read a newspaper when a SAM detonated directly above us. Our Gunner

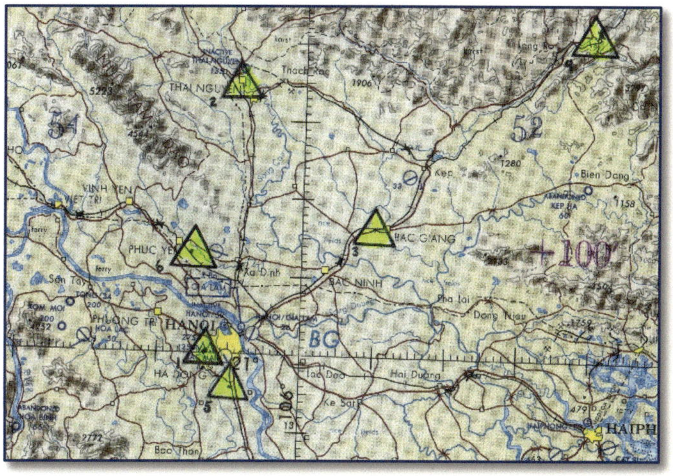

estimated the missile missed us by less than 300' as it flew past. The turn complete, our radio came alive when a MiG tried to attack Aqua Cell - three-minutes ahead of us on the same outbound path. The MiG wasn't successful, and we didn't become his next target. A couple of days later, bomb damage assessment photos showed we destroyed sixty-five buildings in the target area. One report indicated the entire SAM support facility disappeared under the accurate rain of Mark-82s eliminating Hanoi's ability to supply replacement SAMs to the outlying missile sites. When we ended LB-II, there were no military targets left in North Vietnam worth attacking, and we owned the airspace over the country. The North Vietnamese rushed back to the peace talks and agreed to release our POWs. Wise move because the crews were willing to go back and finish them off.

Years later, I was talking to one of the 591 POWs after they were released from North Vietnam. To keep morale up he said they would always tell themselves they would surely be home by Christmas. When that didn't happen, then it would surely be by the next Christmas. For him, his 6th Christmas as a POW was in 1972. When Linebacker II began, and the POWs heard the constant roar of bombs going off, they knew it could only be from the B-52s. The North Vietnamese guards began to shake in fear and tried to stay as close to the POWs as possible as that was the only safe place. As ex-POW George Bud Day said, "When that first rack of bombs fell, everyone started screaming. We knew we were free."

Despite the claims of "carpet bombing" and massive civilian deaths by the liberal left and the U.S. news media, even Hanoi admitted they had only 1,318 civilian casualties from our 700 bombing missions, 10 falling B-52s, the thousand exploding SAM missiles and thousands of rounds of AAA gunfire returning to earth. A testament to the professionalism and skill of all the crews attempting to carry out their missions with minimal collateral damage under very difficult circumstances. My story was not unique.

The true heroes through all of this were the families at home. Despite the constant repeating without question by the US news media of the communist lies about the number of US losses of B-52s (the number claimed by the anti-military US TV commentators sometimes exceeded the number of aircraft DOD owned) and their claims of a terrible US defeat and war criminal status of the crews fighting this "illegal war" and "killing babies", the wives never lost faith. With mail taking up to 10 days to get back and forth to UT and Guam and no internet/emails, they could only wait and pray while trying to shield our children from the ugly media attacks. They are the ones who deserved and should have received the "Welcome Home" greetings!

After Vietnam, I became an instructor pilot and later a flight examiner in the B-52D. Having had my fill of nuclear alert duty, I volunteered for the Air Force's Funded Legal Education program and attended law school at the University of Kansas. Upon graduation, I began my second career in the Judge Advocate General's Corp (JAG). My assignments were in Las Vegas, the Philippines, again at Carswell as the base lawyer, Colorado Springs, and finally, Scott AFB, Illinois. My first assignment as a JAG was at Nellis Air Force Base outside Las Vegas for two years. I was the junior captain in the base legal office and dealt with trials, discharges, claims, contracts, and most everything else. In the Philippines, I worked for the 13th Air Force at Clark Air Force Base. I was a Major

responsible for military justice, and we prosecuted Black Marketing, drugs, fraud, and other major crimes. I then returned to Carswell as the Base Judge Advocate, supervising seven military lawyers and forty reserve lawyers. Finally, as a Lieutenant Colonel, I worked at the Air Force Space Command (AFSC) and North American Aerospace Defense Command (NORAD) in Colorado Springs, where I dealt with some unique environmental work, land exchanges between the military and the State, water rights and whatever else came through the door. Finally, I was sent to Scott Air Force Base near St. Louis to the Military Airlift Command (MAC) as the senior contracts lawyer. That was just prior to Desert Shield/Desert Storm, when the issue of the government exercising its contractual right to call up and control civilian airlines: American, TWA, United, Delta, FedEx, Pan Am, etc. – known as the Civil Reserve Air Fleet (CRAF) and disrupt their normal profitable business in order to transport troops to the Middle East arose. The CRAF program, created by President Eisenhower, had been in existence for 39 years but never used (and thus no "how-to" book) until I got there.

Nona and I recently celebrated 51 great years of marriage with two wonderful children, their super spouses and four fantastic grandchildren. Our son, Jeff, is a Drug Enforcement Agency (DEA) Agent in Detroit, Michigan, and his wife, Jennifer, a nurse. Actor Bradley Cooper portrays Jeff in the 2018 Clint Eastwood Movie, *The Mule*. Our Daughter, Janelle, is an infectious disease medical doctor in the USAF in San Antonio, TX and was recently selected for promotion to Colonel. Her husband, Jeff, is an Army Doctor, also selected for promotion to Colonel. Jeff was prior enlisted Army Ranger and works with Special Forces. He was the Army's nominee in 2017 to be Vice President Pence's doctor (Navy won the coin toss).

I remained on active duty for twenty-five years, retiring as a Lieutenant Colonel. Fortunately, I was not wounded, nor do I suffer from any Vietnam-related disabilities. I received three Distinguished Flying Crosses: 1st - for that sixth mission in Linebacker II, 2nd - for the first five missions, and the 3rd - for "Extraordinary Achievement in Aerial Flight," dated earlier in 1972. I also received the Defense Meritorious Service Medal (MSM), three

Air Force MSMs, 13 Air medals, and many other awards from my six years and 2000 hours flying bombers in the Strategic Air Command, and nineteen years as a Judge Advocate/Staff Judge Advocate.

After I retired from active duty in 1993, I went to work for US Transportation Command (USTRANSCOM also known as USTC), a Joint Services Transportation Command, as a civilian attorney for another twenty years, retiring in April 2014. At USTRANSCOM we supported every military action (troop deployment, medical and casualty evacuation, inflight refueling, redeployment), international emergencies and evacuations, natural disaster responses, plus all the day-to-day transportation needs of the U.S. military and the President. Responsible for international law, civil law, transportation law, ethics and fiscal law in

addition to writing legislation for submission to the US Congress, every day was an adventure for me. The legislation I drafted was to help the command with its mission. We obtained passage by the U.S. Congress of 57 laws – two most "famous" were: 1. The law allowing military and all (not just DOD) government employees to keep their frequent flyer miles from official travel and 2. The law allowing evacuation of family pets with the military family when required to evacuate a foreign country due to civil unrest or disaster. I also became the DOD expert on government provided war risk insurance for airlines supporting military operations where commercial insurance wasn't available or was available only at exorbitant rates. USTC and its components, working with Federal Aviation Administration (FAA) and the Maritime Administration saved taxpayers over $1B in avoided insurance costs. I was very fortunate to work with great people. I retired from USTC in 2014. USTC approved and is currently publishing the book I wrote for them on the government insurance programs.

In 2019, we moved to Florida, where Nona and I spend our time with veterans' activities, photography, travel, and playing the ukulele.

The Distinguished Flying Cross Awards

1st Award:

Captain Dwight A. Moore distinguished himself by heroism while participating in aerial flight as a B-52 Aircraft Commander, 307th Strategic Wing, U-Tapao Airfield, Thailand, on 29 December 1972. On that date, Captain Moore was flying on a high priority mission against a heavily defended target in North Vietnam. Despite the hazards presented by numerous surface-to-air missiles, Captain Moore released his ordinance precisely on target and safely egressed from the high threat area. The outstanding heroism and selfless devotion to duty displayed by Captain Moore reflect great credit upon himself and the United States Air Force.

2nd Award: (First Oak Leaf Cluster)

Captain Dwight A. Moore distinguished himself by extraordinary achievement while participating in aerial flight as a B-52D Aircraft Commander, 43rd Strategic Wing, Andersen Air Force Base, Guam, on 3 June 1972. On that date, while flying on a critical combat mission in Southeast Asia, Captain Moore had to assume the lead of his cell of three aircraft due to a malfunction of the lead aircraft. During the bombing run, Captain Moore was forced to employ emergency bombing procedures due to a further malfunction of an aircraft in his cell. These procedures were completed with great precision despite the presence of a severe tropical storm in the area of the target. As a result of Captain Moore's efforts, all weapons were delivered on the high priority target. The professional competence, aerial skill, and devotion to duty displayed by Captain Moore reflect great credit upon himself and the United States Air Force.

3rd Award: (Second Oak Leaf Cluster)

Captain Dwight A. Moore distinguished himself by heroism while participating in aerial flight as a B-52 Aircraft Commander, 307th Strategic Wing, U-Tapao Airfield, Thailand, from 18 December 1972 to 28 December 1972. During that period, Captain Moore repeatedly attacked heavily defended, vital military targets in North Vietnam. Despite the extreme hazards presented by hostile fighter interceptors, numerous surface-to-air missiles, and high caliber antiaircraft artillery, Captain Moore completely fulfilled his responsibilities as Aircraft Commander without regard for his personal safety and successfully penetrated these defenses, releasing his ordinance precisely on target. The outstanding heroism and selfless devotion to duty displayed by Captain Moore reflect great credit upon himself and the United States Air Force.

B-52 Stratofortress

Mutually Assured Destruction (MAD)

Boeing B-52 Stratofortress with ordnance on display

Throughout the Cold War, B-52s and other US strategic bombers performed airborne alert patrols under code names such as *Head Start*, *Chrome Dome*, *Hard Head*, *Round Robin*, and *Giant Lance*. Bombers loitered at high altitude near the borders of the Soviet Union to provide rapid first strike or retaliation capability in case of nuclear war. These airborne patrols formed one component of the US's nuclear deterrent, which would act to prevent the breakout of a large-scale war between the US and the Soviet Union under the concept of Mutually Assured Destruction (MAD).

Two of the pilots in this book flew Cold War missions in B-52s, 24 hours at a time, with a full load of nuclear bombs waiting for an electronic signal from the White House to unleash nuclear destruction. Their white hair attests to the stress of knowing that any second of that 24 hour flight they might be ordered to execute the nuclear option.

7

John F. Morgan
Captain
United States Air Force

Cold War & Vietnam Era

A young Captain Morgan

It was August of 1966 when I received my Bachelor of Science degree in Civil Engineering from Tri-State University; turned 21 years, and was married to Eileen Karon a couple of weeks later. I started my first job with Armco Steel Corp., Middletown, Ohio, as a future

Sales Engineer selling their manufactured construction products. Around January of 1967, I started receiving draft notices from my Michigan draft board. After considerable thought and visits to the Air Force and Navy recruiters, I decided to "go for it" and see if I could qualify for Officer Candidate School and subsequent Flight School. It was never my burning ambition to be a pilot when I grew up. The thought really became a fixation once I knew I was either going to be drafted (two-year enlistment and Vietnam), or make the best of an opportunity to "be the best" and become an Officer and see if I had the right stuff to fly airplanes. After a thorough examination and evaluation, I was told I had the physical and mental aptitude to qualify for USAF Pilot Training, once I completed Officer Candidate School as a second Lieutenant. After that news, I dropped my application to the Navy and went home to tell Eileen the good news. To say she was not overly excited about me signing up for OCS and Flight School would be highly accurate.

After 53 weeks of Undergraduate Flight Training at Moody AFB, Valdosta, Georgia, I was assigned to McGuire AFB in New Jersey to fly C-141s, a four-engine jet cargo airplane. After about a year in 141s flying all around the world (and bringing home exotic presents for Eileen), I was "asked" to move on to flying AC-119K gunships in South East Asia. I had no idea what the AC-119 K was, so I asked the squadron CO what it was all about. He was quite diplomatic and explained what little he knew about the gunship and concluded the discussion by saying "you will have a very interesting mission." We now had a two year old and six month old daughter and a bunch of Temporary Duty (TDY) for me before leaving for SEA. Finally, about October of 1970, two years after getting my wings from Moody, and less than 14 months in the 141s, I said my good byes to wife, daughters, and family and headed off to war.

I was the Aircraft Commander an AC–119 gunship assigned to the 18th SOS stationed at NKP (Nakon Phenom, Thailand). We had a total of about 20 aircraft split between Danang (Vietnam) and NKP. The old C-119 cargo plane was converted to a gunship by adding two J-85 jet engines to the outboard of the standard reciprocating engines and placing four (4) 7.62 Gatling guns and two 20 mm Vulcan cannons out the left side of the aircraft. They also added a night observation scope (NOS) out the left side and a Forward Looking Infrared Radar (or FLIR) which was invaluable to our mission of Armed (night) Interdiction over the Ho Chi Minh trail. We flew strictly night operations, due to our rather low orbit and slow airspeed needed to place us in the proper firing circle geometry. We flew with a 10-man crew comprised of 5 officers, and 5 enlisted men. Besides our night truck hunting mission, our secondary mission was assisting nighttime troops in contact, (TIC). The airborne command and control center monitored our every mission and would redirect us to any place within reason that was being shelled or in a nighttime combat situation.

John Morgan (4th from right) and his 9-man crew of the AC-119 gunship

Fairchild AC-119 "Shadow" gunship

It is important to note that the following mission summary is not the one for which I received a Distinguished Flying Cross. My DFC was for a mission where we discovered a series of 'movers' on the trail and were able to destroy them with the 20 mm Vulcan cannons while dodging a bit of Anti-Aircraft Artillery (AAA) fire. That story has been published in the National DFC magazine. The mission summarized below was, in my estimation, a much more gratifying and important event than any other of my 180 combat missions.

C-119 with guns blazing

In late June or early July of 1971, we were flying a "dawn patrol" mission over the Ho Chi Minh trail looking for targets of opportunity. We called it the dawn patrol because it was the last mission of the night and usually ended as the sun came up about the time we were landing back at NKP. As we were approaching "Bingo fuel" (minimum fuel for safe return to base) and getting ready to return to base, we received a call from the Airborne Command Control and Coordination center (ABCCC) asking if we were still on site and if we could help a Lima site (Laotian Operating base or site) engaged in a battle with a large force or bad guys threatening to over run their outpost. We were given the coordinates and the contact person's radio call sign and authentication information. I believe it was either Poppy or Rose bud but we will call him Poppy. These folks were part of the Hmong Laotian armed forces fighting Pathet Lao communist forces. As we approached the area, we were unable to visually identify the good guys from the bad guys but could see tracers and mortars coming from different directions. It was still night, but barely. It was squadron policy to be back on the ground before sunrise no matter what the target or circumstances were. We were definitely at a disadvantage flying any sort of daylight operations where AAA or even heavy small arms fire was present. After making contact with the ground commander we were able to pinpoint his location. From there his guidance and our offsets were used to determine the enemy location. The enemy was on a ridgeline approximately 100 to 200 meters from Poppy's hill top Lima site and approaching rapidly, using the last bit of darkness for their assault. Since we would be firing at people within 100 meters of his position, we secured Poppy's permission for a possible short round before our first firing pass. It is always a pucker situation when we needed to shoot that close to friendlies, and every crewmember needed to be at the top of their game. I called for only one of the miniguns (7.62) to be placed on line, and advised Poppy we were rolling in for a short test burst and we would adjust future impacts based on his direction. I rolled into orbit and the co-pilot and flight engineer put me on the proper airspeed and altitude. The first burst was on the East edge of the ridgeline and Poppy whole heartedly approved and asked us to "shoot more and use big gun too." With our FLIR and NOS operators now identifying the target precisely, and the sun starting to crest the horizon, we decided to put all six guns on line and work the entire top of the ridgeline by rocking the wings of the aircraft when we were perpendicular to the ridgeline. With all six guns blazing we put out 30,000 rounds a minute and the 80,000 pound aircraft was shuddering from the percussion as we were physically pushed sideways. We made a total of two passes as we approached the perpendicular points of the ridgeline. When we said good bye to Poppy he was very, very appreciative. We were very glad to help and the entire crew felt like that had truly made a difference and accomplished the mission.

As a side bar we arrived back at NKP quite a bit after sunrise. I knew I was in trouble and would have an "appointment" with the squadron commander after a bit of sleep and before the next night's mission. As it worked out, I was called into his office before our briefing that evening. I explained why I was late returning to NKP and he sat quietly for a minute or two. When he spoke, he said he knew all about the mission and my version was very similar to the one he had heard for General Vang Pao, the commander of Hmong forces in Laos. He also told me to pick out two members of my crew and be prepared to be guests of honor of General Pao at his compound in Laos. Two days later, my copilot Captain Patrick McGillis, and my FLIR operator Lt Colonel Dick Denfield jumped on board an Air America C-123 for the flight to meet General Pao. We had to wear civilian clothes and take no identification with us. After landing we boarded an AA camouflaged helicopter and were flown to the Lima site where we had helped just a few days before. We met some of the fighters and it was a very humbling experience. There were young teens as well as seasoned veterans with a few mortars, M-16s and spotting scopes. That evening we were ferried back to General Pao's headquarters for a great banquet and many toasts to us consisting of straight Scotch whiskey. A great experience.

AC-119 parked on the tarmac

CPT Morgan was awarded the Distinguished Flying Cross in 1971.

Citation:

"The President of the United States takes great pleasure in presenting the Distinguished Flying Cross to Captain John F. Morgan for extraordinary achievement while participating in aerial flight as an AC-119K Pilot in Southeast Asia on 2 November 1971. On that date, Captain Morgan flew an armed reconnaissance mission against heavily defended supply routes being used by hostile forces. After locating several truck convoys, moving under the cover of darkness, Captain Morgan's aircraft made repeated attacks against the convoys despite very intense and accurate anti-aircraft fire. Captain Morgan's superior airmanship resulted in the damage and destruction of several hostile supply vehicles. The professional competence, aerial skill and devotion to duty displayed by Captain Morgan reflect great credit upon himself and the United States Air Force."

Unlike many Vietnam vets, my tour was a "growth experience". Eventually I became Aircraft Commander of a 10-man crew of Officers and Enlisted men, became an Instructor Pilot, flew 180 combat missions, and received a few Air Medals and a DFC. We flew night armed *recon* missions trying to stop the flow of enemy trucks and personnel on the Ho Chi Minh Trail. We also flew air and fire support for ground troops under attack at night. It was an amazing experience and one I will never forget. I learned so much; had so much responsibility entrusted to me as a 25-year-old Captain; and became a different person with different values when I returned home in October of 1971. Unlike many Vietnam vets, I was discharged and sent home to Michigan three days after returning to CONUS from 12 months of war. We landed at Travis AFB in California, were out-processed in three days; I was told to expect protestors/anti-war demonstrators, and to avoid possible physical confrontations with them at my destination, Detroit, where my family was waiting. I was given a dark blue, long raincoat to cover my uniform and advised to not wear my military hat. In other words, hide any appearance of being a military person/officer. It was not the way I had envisioned my return home. I was looking forward to proudly walking off the airplane in Detroit in my dress blues with a new attitude and a new life ahead of me. It just did not work that way in 1971 in the military. I will conclude the military portion of this writing by saying that I was truly enriched by my time and training in the military and my year in South East Asia. I am a better man because of it.

Coming home after 12 months to Eileen and two little girls who did not know me, with no job and no house other than her father's where she had stayed, was unsettling. I quickly found an engineering job in Port Huron, MI. where we settled in and I became domesticated once again. After about a year and a half, I was contacted by my former company, Armco, and was offered a Sales Engineer position in Michigan, which I accepted. I stayed with the company for 37 years.

8

William Rothas
Major
United States Air Force

Cold War and Vietnam

Major Bill Rothas

Bill Rothas had a distinguished US Air Force career during the Cold War* from 1958 through the Vietnam war. He was privileged to fly some of the most advanced aircraft of that era.

Bill was born in Surf City NJ and raised there until the age of 14 when his parents split and he moved with his mother to St Petersburg Florida. Bill signed up for the Navy Reserves at age 14. After high school he enlisted in the Air Force, and began training as an electronic warfare technician. He was transferred from one base to another and eventually discovered the Air Force was beginning a new program that would allow him to begin flight training. Bill took advantage of it, passed all the preliminary testing and reported to Graham Air Base, FL began ground training and began flying T-34A and T-28 A aircraft. (Life was good)

Soon, he was stationed at Bryan Air Base, Texas flying the Lockheed T-33. He graduated from flight school and was commissioned a second Lieutenant in the United States Air Force 12 June 1958—he was 22 years old. He was reassigned to Williams Air Force Base near Mesa Arizona to start flying F-86 fighter jets—life was even better! He met his wife Ellie while assigned to Williams AFB.

From there he was sent to survival school at Stead AFB, then he was selected for B-47 ground school at McConnell AFB Kansas. Bill qualified in the B-47. The Strategic Air Command now had him firmly in their grasp. He flew long missions in the B-47 for two years of the Cold War. In 1958 and 1959 the cold war was heating up due to the space race which the US was not winning. Intelligence, spy gadgets and radar jamming gadgets proliferated. Bill was at the leading edge of those developments.

Following several flying assignments in the US and Europe, Captain Bill Rothas reported to Little Rock AFB as part of an air crew assigned to stand alert for the Cold War. That duty is now world famous for the long hours of strategic readiness and the fleet of aircraft that carried nuclear weapons as part of the deterrent force that maintained nuclear war as a constant threat and made the Russians and any other would-be aggressors pause before they triggered the wrath of America which was at that point the only country that had actually dropped an atomic bomb in anger. He helped maintain the US as a credible deterrent threat.

Bill was still trying to find a way out of SAC. As a master aviator he had experience and skill in various aircraft, and was well versed in many mission profiles, but was given a desk job which did not satisfy his "need for speed." He was offered B-52 school and a follow-on assignment at SAC flying more round the clock missions to deter nuclear war. He declined, telling his commander he wanted out of SAC, and after 2 years at Offutt AFB near Omaha NE he accepted an assignment flying KC-135 airborne refueling tankers. An interesting memory about his flying the KC-135 missions to refuel flights of other aircraft is that the missions were "silent"; there was no radio chatter, the 'tanker' was given a heading, often electronically, took up the heading and the flight of aircraft came up behind him, sequenced themselves according to their SOP, refueled in turn, then turned and went on their way. He remembers how quiet the missions were. Most of the talk was over the intercom within his aircraft.

Then a personal need snuck into his Air Force career, Bill was still struggling to get college credits and complete a college degree. On several occasions people with less flying experience who had college degrees were selected ahead of him for assignments that he wanted even when he was next in line. Bill had two daughters he wanted to spend time with and wanted to complete his college degree. He heard that the EB-66 units were asking for qualified pilots at a location where he would have a chance to work on completing college credits. He reported to Shaw AFB South Carolina for EB-66 training.

Douglas EB-66C Destroyer with uprated electronic counter measures equipment

The EB-66 was an adaptation of the B-66 to carry extensive cameras and electronic warfare equipment like three banks of electronic jammers to confuse and disrupt radar and electronic scanning equipment – especially anti-aircraft missile and gun and tracking radar by blasting the air with white electronic noise that would prevent a radar source from "seeing" any targets. The aircraft could also be fitted with multiple cameras for high-resolution photos -- day or night. Upon completion of EB-66 training he went to he was sent to Takhli, Royal Thai Air base (west of Bangkok) as an EB-66 aircraft commander. His crew for those missions included three to four Electronic Warfare Officers who managed the EW weapons systems and did all the jamming and performed the various intercepts. All he had to do was fly the aircraft to the designated location and watch out for MiG fighters, which occurred several times while flying over Laos along the Ho Chi Minh trail. (Friendly fighter jets would take on the MIG's so he could continue his mission. The EB-66 was not a fighter jet.)

The **Distinguished Flying Cross**: Major Rothas' DFC was not awarded for a single act on one day or one mission, his DFC was awarded for the period from 13 July 1968 to 7 June 1969, a year during which he flew hundreds of hours in the face of hostile antiaircraft and missile fire as he provided radar jamming and electronic anti-aircraft jamming for several hundred missions providing safety for other aircraft in the Southeast Asian Theater. He flew over 800 hours in missions over Cambodia, Laos, and the Ho Chi Minh trail in a variety of poor weather, in dangerous tactical situations that put him and his crew at risk in order to save the lives or guarantee safety of many aircraft flying bombing, reconnaissance, and troop insertion missions into the jungles along the Ho Chi Minh trail. He provided EW support to heavy bombing missions all over South East Asia. He provided anti-missile jamming for numerous of "Arc Light" (B-52 bombing missions) into South Vietnam. There is no way to know how many of us were protected or saved by his actions. Again, it was not for a single day of valorous flying, but for an entire year of extraordinary navigation and flying.

Citation to Accompany the Award Of
The Distinguished Flying Cross
To
Major William J. Rothas

"Major William J. Rothas distinguished himself by extraordinary achievement while participating in aerial flight as an Aircraft Commander at Takhli Royal Thai Air Force Base, Thailand, from 13 July 1968 to 7 June 1969. During this period, Major Rothas flew extremely hazardous missions under the constant threat of hostile ground fire and MIG fighter attack. He frequently used his extensive flying experience and aircraft knowledge to overcome adverse weather conditions and aircraft malfunctions. In spite of these hazards he successfully provided valuable electronic jamming support and threat warnings for B-52 Bomber, fighter bomber and photo reconnaissance flights allowing them to complete their missions in greater safety. The professional competence, aerial skill and devotion to duty displayed by Major Rothas reflect great credit upon himself and the United States Air Force."

Bill submitted this as his most poignant memory of Vietnam:

"Different Missions"

The land appears so peaceful from 30,000 feet, though pockmarked, defoliated, and simmering in the heat.

You're aloft and aloof and thinking of the States, while down below the "Thuds" are due, about to keep their dates.

Guard Channel cracks and come alive, "Mayday, I'm going down, off Eighty nine sixty East-- - Do you read me Crown?"

A pause then a calm, sure voice, "Rog, Scotts Two, we do. We're on our way balls to the wall; we've got two Sandy's too!"

So you sit there in your orbit five miles above the ground; Sun in your face, war far below, nice and safe and sound,

While "Scotch Two", your buddy, Dick, is scrambling to be free,

But there's little time and too many G's. The rest is history.

Your mission's over, success again; it's time to "RTB"

Crown calls on Guard, "Scotch Two, come up!" rather plaintively.

The Officer's club is like death tonight; all eyes are filled with sorrow.

Dick is gone forever now, and I fly again tomorrow.

Notes:

- The "Thuds" were the Republic F-105 Thunderchief supersonic fighter bombers which did most of the early bombing in Vietnam.
- "Scotch Two" was the call sign of Bill's friend, Dick.
- Guard is the international emergency frequency (121.5 & 242.0 megahertz)
- "RTB" was pilot talk for Return to Base.

Bill retired from the Air Force, moved to St Petersburg, then to Monticello near Tallahassee, FL. The Air Force asked him if he wanted to work for the postal service, but he opted into real-estate instead. He did well and enjoyed the real estate work. When the housing market crashed, he was invited again to work with the USPS and he accepted a position as a letter carrier, stayed with USPS until his retirement but always kept his hand in the Real Estate market.

9

Ronald George Schwenkler
Lieutenant Colonel
United States Marine Corps

Cold War and Vietnam

Ronald George Schwenkler

I was born on July 5, 1938 in Chicago, Illinois. My Mother called me a "squib"; that is, a firecracker that goes off (explodes) a day late. Because of my Father's poor health due to crippling arthritis in his hands; we moved to Tucson, Arizona, in early 1942. The move to Arizona did a great deal of good for my Dad because after hours in the sun he became completely cured of his arthritis. In fact, he became a day laborer and one year he was the singles bowling champion of the state of Arizona.

I grew up during the Second World War which gave me a lot of influences and memories. My Dad also worked for Consolidated-Vultee Aircraft (CONVAIR) as an inspector. I saw news reels about the war around the world. Since our home was close to Davis Monthan Air Force Base, I saw many, many different types of airplanes in the sky close by our house. I was always fascinated by airplanes and built many model air planes by myself. Besides interest in aviation and, being boys, we also played as soldiers and Marines imitating our heroes who were actually in combat around the world. Interestingly, with no money we used our ingenuity to make our own toy guns out of left-over building supplies. Of course, I remember all the "bad guys" like Hitler, Mussolini, Tojo, Hirohito, etc.

As a youth I remember seeing the movie "The Flying Leathernecks" starring John Wayne. After that, I remember saying to myself, "Someday I want to be a Marine fighter pilot." More on that later in my story.

After my graduation from an academic, college preparatory high school in Tucson, Arizona, no one gave me any counsel regarding going to college. I acquired a full-time job at Hughes Aircraft as an etched circuitry inspector. Since the "Falcon" missile that was being manufactured by Hughes was not working properly, there was a large lay off of workers, me included. After getting laid off with no other job opportunities available I decided to join the U.S. Navy.

Seaman Recruit Schwenkler, 1957

As a seaman recruit in "Boot Camp" I scored 143 when I took the General Classification Test (GCT). This meant that I was smart enough to face the academic challenges of the US Naval Flight School; like aerodynamics, mathematics and science. So, I applied for Naval flight training.

I was accepted for and arrived at U.S. Naval flight training in Pensacola, Florida, in early 1958 and began my training as a U.S. Naval Aviation Cadet, intending to become a Navy pilot. At the start of training one of the instructors told us to look at our fellow cadets on our left and on our right saying, after that, "during flight training 1/3 of you won't be here because you won't complete or qualify for one reason or another." Following 16 weeks of preflight indoctrination I was sent to NAS Saufley Field, (near modern day Naval Air Station (NAS) Pensacola) FL for primary flight training in T-34B mentor aircraft.

T-34B Mentor

T-28B Trojan

Toward the end of primary flight school, I decided that since most of my friends were either Marine Cadets or Marine Officer students, I applied for and was accepted as a Marine Aviation Cadet.

I soloed on my 12th flight in the T-34B and was sent to the U.S. Naval Air Station Whiting Field, FL for flight training in the T-28B and T-28C aircraft, we conducted formation, night and cross country flying. They were similar planes but the "C" version had a tail hook and shorter propeller which made it suitable for landing on aircraft carriers. We didn't get a chance to carrier qualify in the T-28C due to a shortage of aircraft engines. However, we did go aboard the U.S. Antietam in the back seat with an instructor doing the flying and landings. During that carrier demonstration we lost our first fellow student, Paul King, and his instructor who crashed aft of the carrier.

My next assignment was instrument training in T2V jet aircraft at U.S. Naval Air Station Memphis, Tennessee which was followed by my transfer to advanced flight training at the U.S. Naval Air Station Beeville, Texas, where we were trained in the F9F-8B and F9F-8T. We were given flight training which included formation flying, advanced instrument training, air-to-air gunnery as well as training to land and fly from aircraft carriers. Carrier landings were the final test of our abilities to complete Advanced Flight Training. One of our students was not skilled enough and could not land on the carrier so he was sent back to Beeville where he was dropped from the flight training. He completed his obligation in Navy as an enlisted man.

"Someday I want to be a Marine Corps Fighter pilot"

Ron Schwenkler, age 8

In May 1959, at the age of 20 I was commissioned as a Second Lieutenant in the U.S. Marine Corps Reserve which was immediately followed by receiving my Gold Wings as a Naval Aviator. At that moment I remembered that when I was about 6 years of age, I said I wanted to be a Marine Corps pilot someday. It was kind of like a self-fulfilling prophecy even though I never even thought about it or dwelt upon it all those years.

Following the ceremonies, I was transferred to the U.S. Marine Corps Air Station El Toro, California where I joined Marine Corps Reconnaissance Squadron 3 (VMCJ 3). During my tour there I flew the RF-8A Crusader and the EF-10B Skyknight. Historically photographic reconnaissance aircraft are unarmed. The intelligence gathered from the photos of the target are more important than defending yourself or fighting back.

RF8-A (With Mt Fuji in background)

RF-8 Crusader - Parked

EF-10B Skyknight - In Flight

The RF-8 was the photo reconnaissance version the F-8 fighter and the F-10B that had been a night fighter in the Korean War, (then called the F3D-2) was the electronic countermeasures version of that aircraft. I became qualified to fly both types of these reconnaissance aircraft. Prior to our transfer to the Marine Corps Air Station Iwakuni, Japan, I became carrier qualified in the RF-8A. During this 13-month overseas tour I flew many passive electronic countermeasure missions in the EF-10B off the coasts of Russia and China identifying and locating their radar and navigation sites. A decision was made to resurface the runway at Iwakuni so my squadron was sent to the U.S. Naval Station Cubi Point, in the Philippines. While there, we were still tasked with electronic surveillance of China and Russia in the EF-10B aircraft. The RF-8A's of our squadron were tasked with the almost impossible duty of taking vertical, 15- and 30-degree recon photos of "the entire Philippine coastline." There are so many islands on the Philippine coastline that this nearly impossible task was never completed in our three months in Cubi Point.

In April 1961 while still at Cubi Point, Laos, the Ho Chi Min trail and the Mekong River were of great interest to the United States due to tension in Laos and Vietnam. Since photographic reconnaissance was necessary to discover and prove the transit of goods down the "Trail" the U.S. Navy was tasked to fly these missions off the USS Coral Sea (CVA-43) then stationed in the South China Sea. The Navy needed two more RF-8A's to supplement their RF-8A's. VMCJ 1 was tasked to provide the two photo planes and pilots. The commanding and operations officers decided to give all of the carrier pilots who recently qualified on the U.S. Bonnie Homme Richard one field carrier landing practice (FCLP) session on the U.S. Naval Air Station, Cubi Point, Philippines.

It turned out that I and Bill Stein, the two junior pilots, (two first lieutenants), the youngest ones who were carrier qualified, were selected to supplement the Navy squadron. After preparing some necessary items and checking our fuel load we flew out and landed on the USS Coral Sea. I subsequently flew three TOP SECRET (at that time) photographic recon missions over Vietnam, Thailand and Laos.

Refuel in flight

Landing on the Coral Sea

We would launch from the carrier, rendezvous with an A4, configured with an air refueling pod, refuel in flight, take on a full load of jet fuel and head for South Vietnam. We were tasked to get the photos of "The Trail' from an altitude of 30,000 feet with cameras having telephoto lenses while not making any condensation trails is the sky (That would have shown we were overflying their countries). Immediately after returning and landing on the Coral Sea we were commanded to get directly debriefed by the Admiral of the fleet. Further, we could not talk to ANYONE about our missions; even our squadron commanding officer, after return to our squadron at Cubi Point.

It is interesting to note that aerial photography reached a high level of sophistication as early as 1915 during the First World War. Now this is accomplished by satellites and drones!

After the runway resurfacing at Iwakuni was completed, we returned from Cubi Point to Japan. During our remaining time there I applied for and was accepted for a Regular Commission in the U.S. Marine Corps. Only two lieutenants in our year group were selected for regular commissions; myself and Bill Stein! Following our normal 13-month deployment to Japan we received our orders back to the United States. I was told to report to the U.S. Marine Corps Air Station Cherry Point, NC.

Dating my future wife 1962

It was during this time at Cherry Point that I met my future wife, Morine, on a blind date. She had come to North Carolina with her mother who was visiting an old friend of hers from school in Brooklyn, NY. She looked so athletic and she was so beautiful. Since Morine lived at home with her parents in Brooklyn, NY I would frequently take a cross country flight to the U.S. Naval Air Station Floyd Bennet, NY. On other special occasions, like the Marine Corps Birthday Ball, she would fly down on the airlines to visit me in NC where she would stay with friends of mine Chuck and Carol Dixon. After several years of cross-country dating, we were married on November 30, 1963 at Our Lady, Help of Christians Catholic Church in Brooklyn.

Wedding Day, Nov 30, 1963

As of this writing we have been married for 57 years. We have three adult children and five grandchildren.

After this lengthy tour (1961-1966) at Cherry Point I completed assignments as Assistant Director of the Non-Commissioned Officer Leadership School teaching "The Techniques of Military Instruction." Then I was assigned as Aide-de-Camp to Major General Paul J. Fontana, Commanding General of Combined Marine Corps Bases East and then the Commanding General of the Second Marine Corps Air Station, Cherry Point. During my tour as aide I was promoted to Captain. In 1963, the general was tasked to command of the First Marine Air Wing in South Vietnam at Chu Lai. I declined to accept his invitation to join him going to Vietnam; He said that it was lucky that I said "No" otherwise he would have given me an unsatisfactory in judgement on my fitness report; although I knew it was a joke. I hoped! When my tour as aide was completed, I got the chance to select the squadron of my choice. I chose to join Marine Corps Reconnaissance Squadron 2 (VMCJ 2), also at Cherry Point.

Of course, I immediately returned to flying the RF-8A and the EF-10B. Besides training and keeping current we also flew passive electronic reconnaissance and photographic reconnaissance in and around the island of Cuba. This was the time when the Russians supplied Fidel Castro's Cuba with surface to air missiles. In January 1966 I transitioned to flying the EA-6A (Intruder) and in May of that same year I also became qualified to fly the RF-4B (Phantom II). From that time on I was qualified in all four aircraft and in the RF-8A I re-qualified with day and night carrier landings on the U.S Independence.

Arrival: Danang 1966 with Major General Robertshaw, Major Tucker and myself

In October 1966 we flew six of our EA-6A's from Cherry Point, NC to Danang via Tinker AFB OK, MCAS El Toro, CA, Kaneohe Bay MCAS, Hawaii, Wake Island, Guam, and Cubi Point. Of course, from El Toro to Kaneohe and from Kaneohe to Wake to Guam we inflight refueled from Marine Corps KC-130 F's. Needless to say, those trans-Pacific flights were LONG. In Danang we merged with RF-4B's from MCAS El Toro. The plan was to replace the EF-10B's with the EA6A's and replace the RF-8A's with the RF4B's; however, since there weren't enough EA-6A's the EF-10B's remained. On December 11, 1966 I flew the last RF-8A from Danang to Cubi Point where it transitioned to the U.S. Navy.

Passive electronic sorties were flown to detect and geographically locate enemy radar sites; their search radars and their fire control radars. The active sorties were flown to jam these fire control radars so that their guns could not track our attack aircraft.

On February 27, 1966 (according records in my flight log book), even though my DFC citation mentions May 22, 1966, I was assigned a nighttime active electronic mission over northwest North Vietnam in the Red River valley. Our job was to detect any SAM site activity harassing the strike force and jamming their radar sites, particularly their Fan Song radars seeking to track and fire on our attack aircraft. Our trip up to northern North Vietnam ended up about 25 nautical miles southeast of the Chinese/North Vietnamese boarder in an orbit especially suited to provide electronic active countermeasures against their fire control radars seeking our friendly aircraft attacking from the land and sea. Our mission was to support any and all friendly aircraft during their approach, attack and departure from their target(s).

After a fairly moderate time in our support orbit my Electronic Countermeasures Officer (ECMO), Chief Warrant Officer Bill Fleming detected a Fan Song radar (fire control radar) southeast of our position and almost immediately I saw the surface to air missile (SAM-2) launch towards us in our orbit. I immediately turned towards it and began a high-speed descent towards Mother Earth pretty sure that it could not track me. We descended to an altitude about 1000 feet above the known or suspected mountain heights on a heading for our return to Danang, Vietnam. Every time I began to climb back to altitude the Fan Song would light us up again so we would descend below detection. Of course, since we were spending a relatively long time at low altitudes, we became short of fuel so we diverted to Udorn Air Base in Laos to refuel. After our return to Danang we debriefed as normal and retired uneventfully to our individual "hooches" for needed rest, an adult beverage and sleep.

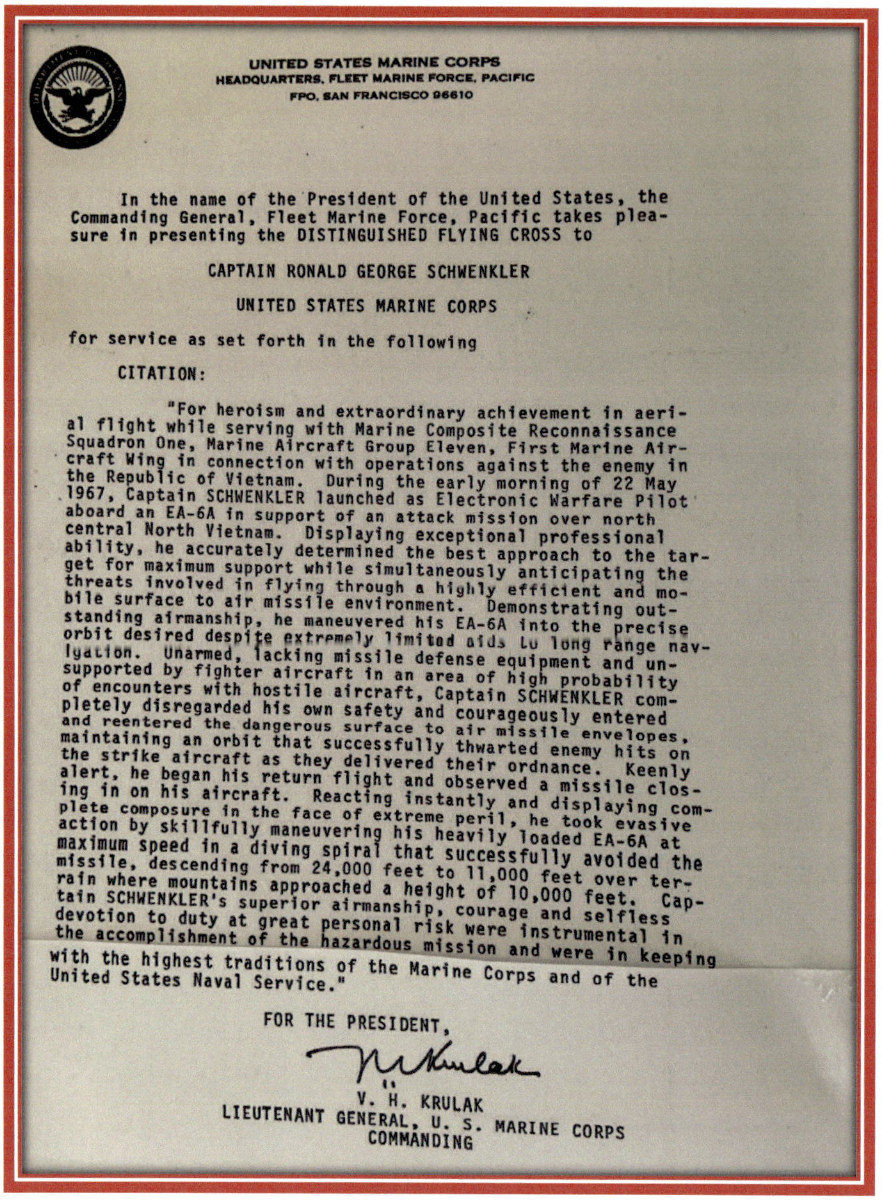

UNITED STATES MARINE CORPS
HEADQUARTERS, FLEET MARINE FORCE, PACIFIC
FPO, SAN FRANCISCO 96610

In the name of the President of the United States, the Commanding General, Fleet Marine Force, Pacific takes pleasure in presenting the DISTINGUISHED FLYING CROSS to

CAPTAIN RONALD GEORGE SCHWENKLER

UNITED STATES MARINE CORPS

for service as set forth in the following

CITATION:

"For heroism and extraordinary achievement in aerial flight while serving with Marine Composite Reconnaissance Squadron One, Marine Aircraft Group Eleven, First Marine Aircraft Wing in connection with operations against the enemy in the Republic of Vietnam. During the early morning of 22 May 1967, Captain SCHWENKLER launched as Electronic Warfare Pilot aboard an EA-6A in support of an attack mission over north central North Vietnam. Displaying exceptional professional ability, he accurately determined the best approach to the target for maximum support while simultaneously anticipating the threats involved in flying through a highly efficient and mobile surface to air missile environment. Demonstrating outstanding airmanship, he maneuvered his EA-6A into the precise orbit desired despite extremely limited aids to long range navigation. Unarmed, lacking missile defense equipment and unsupported by fighter aircraft in an area of high probability of encounters with hostile aircraft, Captain SCHWENKLER completely disregarded his own safety and courageously entered and reentered the dangerous surface to air missile envelopes, maintaining an orbit that successfully thwarted enemy hits on the strike aircraft as they delivered their ordnance. Keenly alert, he began his return flight and observed a missile closing in on his aircraft. Reacting instantly and displaying complete composure in the face of extreme peril, he took evasive action by skillfully maneuvering his heavily loaded EA-6A at maximum speed in a diving spiral that successfully avoided the missile, descending from 24,000 feet to 11,000 feet over terrain where mountains approached a height of 10,000 feet. Captain SCHWENKLER's superior airmanship, courage and selfless devotion to duty at great personal risk were instrumental in the accomplishment of the hazardous mission and were in keeping with the highest traditions of the Marine Corps and of the United States Naval Service."

FOR THE PRESIDENT,

V. H. Krulak

V. H. KRULAK
LIEUTENANT GENERAL, U. S. MARINE CORPS
COMMANDING

Ron Schwenkler DFC Citation signed by Lieutenant General Krulak

There were also many day and night photographic reconnaissance missions flown in the RF4B, one of which became very interesting, to say the least. We were given a target to photograph in the daytime in the vicinity of Vihn, North Vietnam. Every very photographic reconnaissance airplane I have flown has viewfinders; one forward looking, one down looking and one on each side. Of course, the viewfinders are made for locating and photographing the specific target. In order to acquire the specific a photographic plane has to be in straight and level flight. The RF4B engines smoked quite a bit so I developed the following technique: I would extend the speed brakes to slow down the acceleration and put the engines into minimum afterburner so the engines would not smoke so much. That way we would be accelerating throughout the photo run but still say subsonic. As we approached the target, when looking at the downward viewfinder, I could see that their anti-aircraft gunners were shooting at us. I immediately broke off the photo run and headed east toward the sea. Then I realized that we didn't get the photo of our assigned target. I said to my Reconnaissance Systems Officer (RSO), CWO Charlie Offutt: "Charlie we have to go back!" Once over the beach we began the second run toward the target. However, now I couldn't see them shooting at us in the viewfinder. If every fourth round is a tracer, I saw thousands and thousands of tracer rounds; large and small, flying by my cockpit window throughout our photo run. We got our pictures but it really was a miracle that we weren't hit because they had us "nailed"! There was not a scratch on our plane after we landed back at Danang! That was one of the major times that convinced me that I have a great guardian angel whom I named "George"!

"We are not in Vietnam to win."

President Lyndon Johnston, May 1967

As mentioned earlier I intended to remain in the U.S. Marine Corps for the remainder of my career. In May of 1967, at the age of 29, I was selected for and promoted to Major. However, during that time frame our Commander in Chief, The President of the United States, Lyndon Johnson, said that we were not in Vietnam "to win"! I thought about that for about 7-10 days and I resigned my regular commission in the Marine Corps. At that time a letter of resignation was required to be sent to the Secretary of the U.S. Navy, via my Squadron Commander, the Commander of Marine Corps Air Group 11 as well as to the Commanding General of the First Marine Corps Air Wing and the Commanding General of the Fleet Marine Corps Pacific. Of course, this resignation also meant personal interviews with each of the first three commanders mentioned above. During my previous tour at the U.S. Marine Corps Air Station Cherry Point while an aide-de-camp I had met Major General Norman Anderson who was then the current Wing Commander at Danang.

He reminded me that I had a permanent career with the Marine Corps with a guarantee of 22 years of active duty. I told him the I could no longer keep flying these missions if we were not there to win. Further, I told him that when I was a boy, I played marbles to win more marbles and to play all sports with a desire to win not to just break even. I also let him know that eleven (11) of my Squadron Mates had been lost! I just couldn't keep flying these futile missions although I did complete my tour in Vietnam in November of 1967.

During my thirteen months in Vietnam I flew 283 combat sorties, flying these four aircraft: RF-4B, EA-6A, RF-8A and EF-10B. These sorties were about 50/50 photographic, passive electronic and active electronic reconnaissance.

For my next set of orders, I was sent back to Cherry Point, NC where I again flew in VMCJ 2 until the Air Group commander found out that I had resigned my commission. As a result, he removed from the squadron and transferred me to a staff job in the Air Group headquarters as an assistant supply officer. I was not permitted to leave active duty until I waited for one year after my date of my letter of resignation. I was finally released from active duty and spent the remainder of my Marine Corps career as a reserve officer flying the A4B and A4C type aircraft. I was also privileged to become the Commander of Marine Corps Headquarters and Maintenance Squadron-43 for two years at the U.S. Naval Air Station Willow Grove PA. During that tour I was promoted to Lieutenant Colonel. My Marine Corps Reserve career was completed as the Air Officer in a staff group based on Long Island, NY.

I retired from the Marine Corps Reserves as a Lieutenant Colonel with the following decorations: The Distinguished Flying Cross; 13 Strike/Flight Air Medals; one Single Mission Air Medal, the Navy Commendation Medal with combat V; the Presidential Unit Citation and the usual theater medals/ribbons.

Decorations and Ribbons

Following my release from active duty in the Marine Corps I applied for and was accepted as a pilot for Pan American World Airlines based at John F. Kennedy Airport in New York City. At first, I was assigned as a replacement copilot in the Boeing 707 for flights lasting over 9 hours, block to block that means the time from the start of taxi for takeoff until the engines are shut down at the destination airport. Depending upon the Captain in command I would get an occasional take-off and landing in various international airports in Latin America. After a few months of that type of flying I was assigned Flight Navigator Training. Neither GPS or Inertial navigation systems (INS) were available so they were required to have flight navigators on board many flights, especially those transiting Central and South America. After ground school we had to fly 100 hours with a qualified navigator to become proficient in various forms of navigation. My flight check rides were from New York to Lisbon at night and from Lisbon to San Juan, Puerto Rico in the daytime. During these flights I was required to navigate the airplane with these forms of navigation: Pressure pattern flying, LORAN, CONSOLAN and affirming our position by using a sextant shooting the stars, moon and sun, all while traveling at Mach .8 or 480 nautical miles per hour. There were two highly qualified navigators who gave me my check ride: The head of flight FAA Navigation in New York and the Head of the Navigation School with Pan American. Although I thought that I had a lifelong career with Pan Am, it lasted only 18 months when several hundred of us junior pilots were put on furlough (laid off) for an unknown period of time. My furlough date was New Year's Eve 1969!

Since there were no flying jobs readily available, I accepted a job with Reuben H. Donnelly that made the Yellow Pages for New York Telephone. With Donnelly I supervised about 135 women in production sales and in customer assistance. After about a year and a half there I was given a job as an executive pilot with Pan American World Airways Business Jets Division flying Falcon 20 executive type aircraft manufactured by Avion Marcel Dassault in France. Our main mission was to sell Falcon 20 and later, when developed and manufactured, the Falcon 10. We flew demonstration flights in attempts to convince customers to purchase their own aircraft; post production test and acceptance flights from the French factories in Bordeaux and Marseille, France; experimental flight test of and certification of new auto pilot systems and, of course, post production test fights on airplanes purchased by and outfitted for the companies that bought a particular aircraft. I flew for Pan Am Business Jets/Falcon Jet Corporation until 1977.

In March 1977 I was hired by the Wm. Wrigley Jr. Company who had purchased a Falcon 10 aircraft. Another pilot, a mechanic and myself moved to the Chicago, IL, area and were based at what was then called Palwaukee Airport, about 10 miles north of O'Hare Airport. After several years, Wrigley's decided to purchase a Falcon 50, three engine aircraft. A few years later we purchased a Gulfstream 2B; later on, a Gulfstream 4; then two Gulfstream 4's. Finally, we purchased a Global Express jet manufactured in Montreal by Bombardier Aircraft. Since Wrigley's chewing gum was sold all over the world and manufactured in numerous countries. We ended up flying all over the world. I have flown around the world three times for Wrigley Company and flew for them for 24 years.

I ended up with over 14,000 flight hours which is over one year of my life, 24 hours per day, in the air! I never had to run away from an airplane that I flew; thanks to my guardian angel!

While I was busy gallivanting around the world as the CAPTAIN of my ship; my wife, Morine, from the day we were married, was the GENERAL of our ship at home; keeping us united, stable, healthy and solid. Most of this could not have happened without her!

Ronald and Morine Schwenkler, The Villages, FL

Ron's 80th Birthday with Family in Northbrook, IL

10

Dennis (Denny) P. Sharon
Colonel
United States Air Force (Ret)

Cold War, Vietnam

Cadet Denny P Sharon, West Point (1955)

Dennis P. Sharon began his military career as a West Point cadet in 1954. 'Denny' was part of the last class at West Point that allowed graduating cadets to choose the Air Force for their active service career.

"After four years of marching, I was more than ready to try flying. I entered the US Air Force in 1958, went to primary pilot training at Bartow Air Base, FL. flying T-34s, and T-37s; and then went to Basic Flight School at Laredo Air Force Base, Texas where I flew T-33's. Next I transitioned into F-100 SuperSabers, went to Luke Air Force Base in Arizona and Nellis Air Force Base, Las Vegas, Nevada. My next assignment was with the 429th tactical fighter base at Cannon Air Force Base New Mexico flying F 100s. I made a quick switch to F-101's at Bentwaters Royal Air Force Base in the United Kingdom where I was stationed for three years. My next check out was on the F-4, followed by three years as an instructor pilot at Davis Monthan Air Force Base, Tucson, Arizona".

For the uninformed reader, Dennis was fast-tracked on the finest fighter aircraft in our US inventory, which speaks highly of his skill as an aviator. He was the best of the best in his time.

Captain Denny Sharon

Beechcraft T-34 Mentor

Lockheed T-33 Shooting Star

Cessna T-37 Tweet

North American F-100 Super Sabre

McDonnell F-101 Voodoo

McDonnell Douglas F-4 Phantom II

The next step in his military career was his first combat tour flying F-4D's from the Royal Thai Air Force base at Udorn, Thailand. Most of those missions, although based in Thailand were direct combat missions flown into Vietnamese airspace or North Vietnamese airspace with overflights of Laos. (Check a map; he was flying across Thailand, Laos and Cambodia to reach his assigned target areas in North and South Vietnam—very serious flying across several Air Defense Identification Zones!

During the next few months, I received my three Distinguished Flying Crosses for missions flown in June and July 1968. Details are a little fuzzy now; one involved a heartwarming rescue of a downed Marine pilot. The second involved quieting some guns on the ground that were brazen enough to shoot at my wing man; and the third involved a night reconnaissance mission when the bad guys tried a nighttime run south on the Ho Chi Minh Trail. All three missions were exciting but none was as exciting as my second tour at the end of the war, and Christmas bombing of 1973 when we shut down the war, I went back to Seymour Johnson Air Force Base, North Carolina and took command of the 334th Fighter Squadron (one of the original Eagle squadrons from World War II). From there I went on to the National War College in Washington DC, then had a tour as the chief of Air Force liaison to the US Senate where I met Senator Barry Goldwater whom I later worked for.

In July 1977 I took over as director of operations for the 56th Fighter Wing at MacDill AFB, Florida.

Sen. Barry Goldwater Came down to MacDill to fly in my last official Air Force flight in the fall of 1978. I followed Senator Goldwater back to Washington DC as the Minority Counsel of the Senate Intelligence Committee, then as his special assistant for Military, Commercial Aviation, and Space Matters—a position that kept me involved in aviation including the evaluation flights in the F–15, F–18, AV– 8 Harrier, FB –111, A H– 64 Apache, The British Hawk trainer and a couple of Learjets.

McDonnell Douglas F-15 Eagle

McDonnell Douglas F/A-18 Hornet

General Dynamics F-111B Aardvark

McDonnell Douglas AV-8B Harrier II

Boeing AH-64 Apache

McDonnell Douglas T-45 Goshawk, a highly modified version of the BAE Hawk land-based training jet aircraft

After my job in the Senate I joined the McDonald-Douglas Corporation, first as legislative director then eventually as the Vice President Eastern region for the corporation in Washington. I retired again 1996 and did consulting for a few years.

My smartest decision of all was marrying Shirley in 1958. We have a daughter, two sons and three granddaughters. The boys were both army officers. The oldest retired from the Army with 26 years' service, the younger is a Colonel in the Army reserves and is employed at FEMA (Federal Emergency Management Agency). The oldest granddaughter continues the aviation tradition and is employed by Boeing in Mesa Arizona in logistics on the latest version of the Apache helicopter.

What a ride it has been!

Citations to Accompany the Award of The Distinguished Flying Cross to Dennis Sharon

First Award:

"Capt. Dennis P. Sharon distinguished himself by extraordinary achievement while participating in aerial flight as an F4–D aircraft commander over North Vietnam on 19 June 1968. On that date, while leading a flight of two F-4 D's on a night armed reconnaissance mission along a heavily defended and extensively used infiltration route, Captain Sharon executed an aggressive and skillful attack on a hostile defensive position that was firing at his wingman. The explosions resulting from his ordinance impacted the gun complex ammunition dump spread fires throughout a co-located storage and transshipment depot causing extensive damage to a key facility in the North Vietnamese infiltration network. The professional competence, aerial skill, and devotion to duty displayed by Capt. Sharon reflect great credit upon himself and United States Air Force."

Second Award:

"Capt. Dennis P. Sharon distinguished himself by extraordinary achievement fall participating in aerial flight as an F– 4D Aircraft commander over North Vietnam on 13 July 1968. On that date, Capt. Sharon coolly and skillfully overcame the hazards of poor weather and heavy antiaircraft artillery fire to strike a truck park and storage area, destroying active hostile defensive positions and a large number of stores with superior ordnance delivery technique. His aggressive skill in the face of withering hostile defensive ground fire spearheaded further strikes in the lucrative target area and significantly contributed to the strategic interdiction program then underway in the theater of operations. The professional competence, aerial skills, and devotion to duty displayed by Capt. Sharon reflects great credit upon himself and United States Air Force."

Third Award:

"Capt. Dennis P. Sharon distinguished himself by extraordinary achievement while participating in aerial flight has an F–4D aircraft commander over North Vietnam on 26 July 1968. On that date Captain Sharon attacked and destroyed four antiaircraft guns which endangered a rescue effort for a downed Marine pilot. Though downed in an extremely dangerous area, the Marine pilot was picked up soon thereafter due in part to the extraordinary achievement of Captain Sharon. The professional competence, aerial skill, and devotion to duty displayed by Capt. Sharon reflects great credit upon himself and United States Air Force."

Colonel Dennis Sharon, USAF (Ret)

SR-71 Blackbird

In his book **Sled Driver**, SR-71 Blackbird pilot Brian Shul writes: "I'll always remember a certain radio exchange that occurred one day as Walt (my backseater) and I were screaming across Southern California 13 miles high. We were monitoring various radio transmissions from other aircraft as we entered Los Angeles airspace. Though they didn't really control us, they did monitor our movement across their scope.

I heard a Cessna ask for a readout of its groundspeed."90 knots" Center replied. Moments later, a Twin Beech required the same. "120 knots," Center answered. We weren't the only ones proud of our groundspeed that day as almost instantly an F-18 smugly transmitted, "Ah, Center, Dusty 52 requests ground speed readout. There was a slight pause, then the response, "525 knots on the ground, Dusty."

Another silent pause. As I was thinking to myself how ripe a situation this was, I heard a familiar click of a radio transmission coming from my backseater. It was at that precise moment I realized Walt and I had become a real crew, for we were both thinking in unison. "Center, Aspen 20, you got a groundspeed readout for us?" There was a longer than normal pause.

"Aspen, I show 1,742 knots." No further inquiries were heard on that frequency.

11

Dean R. Vetter
Lieutenant Colonel
United States Air Force (Ret)

Cold War & Vietnam

Young Captain Dean R. Vetter, USAF

Why My Crew and I Were Awarded Two DFCs

By Dean R. Vetter, USAF, Ret

I was drafted in November of 1961, when the Berlin Wall went up. I retired in March of 1986, when negotiations began to bring the Wall down!

I joined the USAF to avoid the Army and spent nearly four years in Hawaii, where I finished college. I was then sent to OTS, in Texas and NAV-Bomb training at Mather AFB, near Sacramento, CA. Then to B-52 "initiation" at Castle AFB, CA, just South of Sacramento.

I spent a little time at Homestead AFB, FL before SAC was "shipped" to Wright Patterson AFB, OH, with a few of us (including me) going to Barksdale AFB, near Shreveport, LA in 1968, where we stayed for nine years.

The "D" models had been modified to enlarge the bomb bays and were the only B-52s involved in SEA. We first went "TDY" (Temporary Duty) in early 1970 and again in late 1971, "Merry Christmas!" We were only home for a few weeks, when "everyone" was sent back to Guam.

Boeing B-52G Stratofortress

When we went back in early 1972 it was for Nixon's big build up to get the North Vietnamese to the peace table. This is when nearly all the G models went to Guam. Our crew was ferried to Seymour-Johnson AFB, NC, to fly a cell of B-52s to Guam.

We could not be "TDY" for more than six months, so we were sent home for 30 days of R&R, in mid-year. This is when I was handed several awards and decorations, with a simple "Thank You!" Previous returning crew members had been celebrated on the local TV News.

Returning after the R&R, we flew many more missions on the way to a (my) total of 180 missions.

A Typical Buff Mission Day in Southeast Asia, Circa 1971

By Dean Vetter, Navigator, USAF, Ret

B-52 Bomber Operations were around the clock. The D models had been modified in 1965 so their two bomb bays could carry a larger bomb load.

When Nixon added B-52 G models to the war (see "Spy In The Hanoi Hilton" on YouTube), as 1972 got started, they were stationed on Guam, at Anderson AFB, all year. The D's then flew only out of Thailand and Okinawa, with a few still on Guam.

I had flown D's in 1970 and 1971. The Thailand missions were four hours, with another four hours (two hours at each end) of briefings, loading, de-briefings, etc. (12 hours if you flew out of Okinawa).

Flying out of Guam, however, was a 13-hour mission with refueling over the Philippines. The two hours at either end, for "admin," made it a 17-hour day! We then had a "day" off (24 hours) before flying again. After doing this 180 times, to Laos, Cambodia and N. Vietnam, it still never became "routine."

In Guam, the runway went down and then back up which slowed, somewhat, the B-52s progress toward the "rotate" speed point. The runway ended at a 400-foot cliff. It was somewhat like launching from a carrier. B-52s weigh half a million pounds at takeoff, fully loaded. Half of that is fuel!

If everything was not up to the required mathematical formulas as calculated by scheduling, flight planning and mission briefings, the aircraft would sink (lose altitude) when "launched" off the cliff. If that were the case, I would immediately, upon the pilot's command, make the necessary equipment connections (from memory) and press the bomb release buttons.

The plane would then lose weight and gain altitude. It never happened to me (we did have an "almost" one night but it did not happen. I'm told it is a "hairy" moment. There is, however, one B-52 at the bottom of that cliff.

Older pilots and RNs (Radar Navigators) tend to sleep on long missions and after refueling, while the co-pilot and navigator guide the aircraft toward the target area. We never could make a bomb run with them asleep. They always woke up at the IP (Initial Point).

As a side story (you have to maintain your sanity somehow during the droning, boring hours); my Radar Navigator (RN) and I could both do Donald Duck impressions, so, one day we did the entire bomb run check list with both of us reading and responding as Donald. I think that may have been a mission out of Thailand. These readings are recorded but we NEVER heard a word about it! Years later, a friend who worked in mission analysis at that time, told me they had a good laugh and he hid the tape. It's been said and it's true, that a bomb run is comparable to sex... It's a long road to a specific goal, the excitement mounts, well, you get the idea. We even smoked afterward, with the oxygen off of course.

On the way home, the crews were repeatedly warned to maintain the planned speed (don't rush home) as they would, otherwise, overtake the slower D models in front of them, returning from an earlier mission.

After landing in Guam, I had to go out the "back door" through the wheel well and into the bomb bay to verify it was empty. I could then jump down to the tarmac and see "who buys the beer." We flew as a crew of six and would place initials in each 1/6th of the front tire. Whoever's initial was lowest to the ground, at landing, bought the beer that night.

After mission and maintenance debriefings, we could head for the chilidogs and beer. We were young, but after 17 hours, you could sleep well, no matter what you ate or drank! Then we could head to our "room", divided in two by lockers, three beds on each side and one bathroom to share. We'd have a day off and do it all over again!

Although it was quite confusing where one day left off and the other began. It never seemed like we had a day off. When we did have more free time, we'd hop into our "Guam Bomb" (an old junker a crew would buy as one) and go around the island one way and then the other way the next time.

We could dine at one of the few local restaurants, the officers club or at the Navy Base Club. On Sunday, at the Navy base, you could grill your own steak out on the BBQ. The Anderson Officers Club was usually crowded as most of us didn't like to do anything in our room but sleep.

It was very different when Nixon ordered the bombing of Hanoi, in late December of 1972, officially known as "Linebacker II". Missions were no longer routine. Now we had to deal with the fact not everybody would be returning. All the base facilities ran 24/7 as we did. When your friends left for North Vietnam, you waited up, at the club, until they (hopefully) returned. No one waited alone.

This is what a B-52 crew did for all of 1972 and early 1973, until the POWs came home, several of who had been in one of the 17 B-52s downed in December's "Linebacker II's" successful campaign.

We were all awarded DFCs for making those bombing raids and "participating in one of the most massive conventional bombing raids ever amassed in the history of aerial warfare".

Washington then authorized "Linebacker II", which sent B-52s North for 12 days, in late Dec, 1972 and brought the North to the Paris Peace talks, resulting in the release of the POWs about three months later. Many of those returning POWs were B-52 crewmembers. And far too many had their names added to the "Wall" (See "The Wall and all…"a memoir by Dean Vetter). One POW was a B-52 "Nav" who elected to go to the Seminary and is now the retired Chaplain Rev Col, Robert Certain. You can see his story on "YouTube". He also authored "The Unchained Eagle". Robert's crew was the first B-52 shot down that December night. I am trying to get him to speak in The Villages, FL, in the near future. He is willing but scheduling is difficult.

SAC lost 17 B-52s on the first two nights of bombing, until the tactics were changed. My crew flew over Haiphong on Dec 19th. Sorry to say we missed "Jane" that night! It is very interesting to contemplate how I prepared to bail out of a perfectly good airplane, that night. We just followed our ingrained training and left our wallets, name tags, etc. to stay with the plane if we jumped. You check your Ox bottle, tighten your chute and helmet, reach for the arming lever and think; "What the Hell am I doing"?

I'll always remember the midnight Mass that Christmas Eve. A Chaplain conducted it on the beach. About 25 of us attended, including a few wives whose husbands were "MIA". The Chaplain then invited us all back to his base housing, for breakfast. That was one of the best meals of my life and I don't even know his name.

"They" had a Christmas bombing pause, which allowed the North to "reload" more "SAMs." Of course, I am talking about the G models flying out of Guam, the "Ds" participated from Thailand and maybe Okinawa, I don't recall.

First Distinguished Flying Cross Award Citation.

"The President of the United States takes great pleasure in presenting the Distinguished Flying Cross to Captain Dean R. Vetter for extraordinary achievement while participating in aerial flight as a B-52 Navigator near Hanoi, North Vietnam, on 19 December 1972. On that date, Captain Vetter participated in one of the most massive, conventional bombing raids ever amassed in the history of aerial warfare. While attacking the target, Captain Vetter's aircraft came under heavy and intense fire from surface-to-air missiles and anti-aircraft artillery. As a result of his heroic actions and professional skill, the assigned target was attacked, destroyed, and his aircraft and crew returned safely in spite of the devastating air defense directed against them. The professional competence, aerial skill, and devotion to duty displayed by Captain Vetter reflect great credit upon himself and the United States Air Force."

Our second DFCs say they are for flying more missions in North Vietnam, but they were not to "Downtown". This is also where the total number of missions was involved in receiving the Award.

Second Distinguished Flying Cross Award Citation.

"The President of the United States takes great pleasure in presenting the Distinguished Flying Cross to Captain Dean R. Vetter for extraordinary achievement while participating in aerial flight as a B-52 Radar Navigator north of the Demilitarized Zone in North Vietnam on 8 January 1973. On that date, Captain Vetter's crew was tasked with leading a cell of B-52s against a heavily defended North Vietnamese logistic storage and troop staging area. Despite numerous thunderstorms throughout the target area and the threat of hostile anti-aircraft artillery and surface-to-air missiles, Captain Vetter conducted the bombing run in a precise and professional manner, resulting in an effective weapon release on the target. The professional competence, aerial skill, and devotion to duty displayed by Captain Vetter reflect great credit upon himself and the United States Air Force."

In Jan '73 I was able to go to "SOS" (Squadron Officer's School) at Maxwell AFB, AL. I was still there when they conducted "Operation Update" for all the POWs, but mostly for those who had been held for years. When we left Maxwell, they told us to remember the time difference and to set our watches ahead, "one hundred years!"

Unbelievably, we went back to Southeast Asia, in model Ds, one last time, in mid-1973! Now I was an RN (Radar Navigator – Bombardier) and I had several new crewmembers that were among the "uninitiated," to "babysit." So, I'm not complaining (bitching maybe), but I was in SEA for 25 months out of a 3 ½ year period, or 58% of the time.

Now home to finish a Master's Degree and move to South Dakota to be a part of the SAC "Looking Glass" airborne command post. Then on to Southern CA to be on the numbered AF Battle Staff where I got to watch a few days of Clint Eastwood filming "Firefox" in our Command Post.

I moved to Wright Patterson Air Force Base (WPAFB), OH in 1982, where I served in a SAC detachment as a liaison with Military contractors, updating the BUFF's electronics. When I retired, in 1986, I earned an MBA, on the GI Bill (use it or lose it). After graduation, I worked with a Defense Contractor dealing with training and trainers for F-15 mechanics, crew chiefs.

I have now lived in Paradise for the entire 21st Century!

Dean R. Vetter May 29, 2015

12

Richard W. Yocca
Captain
United States Air Force
Cold War & Vietnam War

A young Dick Yocca

Richard Yocca graduated in his Air Force blue uniform from Reserve Officer Training Corps (ROTC) at Penn State, September 1967 and entered active duty as a 2nd Lieutenant of the US Air Force on 7 December 1967 (Pearl Harbor day). He went through primary reconnaissance and flight training and was qualified as a Reconnaissance Systems Officer or RSO in the McDonald-Douglas RF-4C Phantom. At that time the F-4 was the latest and greatest fighter jet in our inventory. The RF-4C version of the F-4 was designed strictly for reconnaissance with no weaponry on board. This was intended to keep the aircrews' attention focused on gathering intelligence, not trying to shoot at the enemy.

McDonnell Douglas RF-4C Phantom

I was qualified in the RF 4C, (the C model had "no guns, or armor, it had reconnaissance cameras and electronics. We were 'Alone, Unarmed, and Unafraid'.") Our only real defense other than electronics was to light the afterburners and go faster than anything else around us. After qualification I enjoyed assignments at Bergstrom AFB, Texas, various training areas in West Texas, and the Royal Air Force Base, Alconbury, UK.

I arrived in Udorn Royal Thai Air Force Base in Thailand in August 1972, a very interesting day. Our hooches were right next to the officers' club and I had arrived on the day Captain Richie got his fifth MiG. He was an official "Ace", the first Air Force Ace of the war in South East Asia. He was a member of "Triple Nickle"—the 555th fighter squadron—and had shot down five Russian made MiG fighters, so there was a lot going on in the club that night.

Captain Yocca and Major Richie in Udorn, Thailand

My typical missions in Thailand were reconnaissance flights into Laos, Cambodia and North Vietnam. Usually we had fighter escorts and the missions were two to four hours in duration.

The day I earned my first Distinguished Flying Cross was an interesting day. December 27th was the day of the Bob Hope Christmas show at Udorn. We launched at about the time the Bob Hope show was starting up, and if you are able to hear a re-broadcast of that show the jet fighters in the background roaring down the runway were us. Our mission that day was a "Linebacker" mission. For those who don't remember, Linebacker missions were a big deal characterized by multiple aircraft assets at several altitudes and the coordination of multiple air assets over an extended area. We had reconnaissance, and fighter jets and our target was the Pol Domer bridge in Hanoi. We departed Udorn with two reconnaissance RF-4C's and two F-4 fighter escorts. We were not alone, and there were several kinds of support aircraft at various levels above us including the airborne command, control and coordination aircraft. Our mission that day was Bomb Damage Assessment or BDA of the several bridges in the heavily defended Hanoi area including the Pol Domer bridge in Hanoi. We knew there was a high probability we would be shot at. The North Vietnamese had a large number of ground-to-air and anti-aircraft assets in the area around Hanoi.

Our four aircraft left Udorn en route to an airborne tanker in the skies over Laos. After refueling we dropped our external fuel tanks, and headed for Hanoi. Our RF-4C's had compartments capable of containing either night photo cartridges, or chaff (canisters of metallic flakes, plastic and aluminum foil) to confuse and distract radar seeking missiles and disrupt radar. About 50 miles out we started looking for the targets. To get good photography we had to be on a precise azimuth and a set altitude (e.g. 5000 feet), with wings level at about 480 knots. As we were trying to set up our approach, I saw our electronic countermeasures scope light up meaning someone had 'painted' one of us by radar, I also remember a big surface to air missile coming up from our right front, and two silver glints high above us in the sky. There was flak from antiaircraft fire in the area, we knew there was also ground fire from small arms and antiaircraft machine guns, but we would not see the tracers until they were above us. We made our photo run. I turned on all the cameras, we kept the wings level, took the pictures, and turned around heading for home. We never forgot that our mission was not accomplished until after we got home and delivered the film to the technicians. On our way out of the target area we found that the glints in the sky were in fact MiGs. We went to full afterburners to get out of there. Our escorts did good, they shielded us from the MiG's and we got away clean. Two other F-4's were lost that day, shot down by MiGs. There was a lot of hostile fire and there were a lot of airplanes in the air that day. With all the small arms fire and antiaircraft fire and at least four MiGs, we are lucky we did not sustain more friendly losses. We made it back to our midair refueling point, refueled, and returned to Udorn without further incident. Although…we did go down low level on the way back and buzzed a village just to razz the people on the ground a little, but that is another story.

Captain Richard Yocca's First Distinguished Flying Cross

Citation:

> "The President of the United States takes great pleasure in presenting the Distinguished Flying Cross to Captain Richard W. Yocca for heroism while participating in aerial flight as an RF-4C Weapon Systems Officer near Hanoi, North Vietnam, on 27 December 1972. On that date, Captain Yocca led a high priority mission assigned to obtain bomb damage assessment of seven targets in the heavily defended Hanoi area. Undeterred by intense hostile reaction, Captain Yocca achieved complete coverage of all targets and provided higher headquarters with vital intelligence data. The outstanding heroism and selfless devotion to duty displayed by Captain Yocca reflect great credit upon himself and the United States Air force."

My second DFC was awarded based on an accumulation of missions again in the RF-4C.

One of our missions was to reconnoiter the trails leading out of Vietnam into Laos. Some of the missions were to cover the trails in southern Laos, others covered the trails in northern Laos near the plain of jars—"bullwhip missions". We developed a certain amount of expertise for various areas of Laos, and this was a bullwhip mission. We were to go up to the Plain of jars (Plaine du Jares) using a new type of forward-looking infrared radar, which could seek out "hotspots". Those could be heat sources like engines in a truck or jeep, often they could be several people walking in the jungle at night. With these missions you never knew when you would come under fire from the enemy. Usually it was small arms fire. But there were MiGs in the air over Laos also. We flew 169 combat sorties over Southeast Asia, 31 of which were flown over North Vietnam. For that I got a 2nd Distinguished Flying Cross. The citation lists only a mission near Ban Ban, Laos on 6 May 1973. It was a typical mission, but we flew 169 of them often into enemy fire and always into unknown conditions which is why I was presented with 14 awards of the Air Medal and my second Distinguished Flying Cross.

Citation:

> "The President of the United States takes great pleasure in presenting the Distinguished Flying Cross to Captain Richard W. Yocca for extraordinary achievement while participating in aerial flight as an RF-4C Weapon Systems Officer near Ban Ban, Laos on 6 May 1973. On that date, Captain Yocca planned and flew a visual reconnaissance mission to determine the location of suspected enemy gun positions along the roads leading out of North Vietnam. Despite anti-aircraft ground fire, Captain Yocca obtained excellent photography providing higher headquarters with much needed intelligence information. The professional competence, aerial skill, and devotion to duty displayed by Captain Yocca reflect great credit upon himself and the United States Air force."

Then there was the event when I qualified for induction into the Red River Valley Fighter Pilots Association. The criteria were that you had to fly over the Red River North at any altitude and make it home safe. The Red River North flows from China through Laos and North Vietnam along some of the most beautiful country-side in the world, but it was hostile territory bristling with anti-aircraft guns and missile emplacements on the ground, patrolled by MiGs in the air, and thousands of people with guns on the ground and hillsides.

We made it over the Red River North and back without major incident, all of us got home safe.

Aerial view of the Red River Valley, North Vietnam

One of my most poignant memories was an experience in Thailand. There was a full moon, and as I looked up at the moon, I saw flights of B-52s going by at 33,000 feet silhouetted against a full moon. It would have made a remarkable photograph and it is a remarkable memory.

After 11 years in the Air Force I returned to civilian life. I used the GI Bill to get an education in Computer Programming and worked for the State of Pennsylvania, eventually becoming a Computer Network Administrator, I was at the leading edge of modern computer network engineering.

Extraction Under Fire

I will depart as soon as I have all your troops on board.

A typical Huey extraction mission of the Vietnam War

SECTION 4: VIETNAM

At the time it ended the Vietnam War was the longest conflict in American history and the most unpopular American war of the 20th century. It resulted in nearly 60,000 American deaths and in an estimated 2 million Vietnamese deaths. Even today, many Americans still ask whether the American effort in Vietnam was a blunder, a necessary war, or whether it was a noble cause, or an idealistic, if failed, effort to protect the South Vietnamese from totalitarian government.

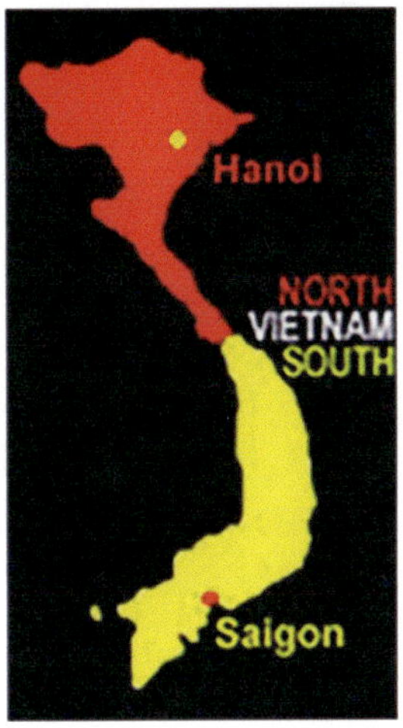

Between 1945 and 1954, the Vietnamese waged an anti-colonial war against France, which received $2.6 billion in financial support from the United States. The French defeat at the Dien Bien Phu was followed by a peace conference in Geneva. As a result of the conference, Laos, Cambodia, and Vietnam received their independence, and Vietnam was temporarily divided between an anti-Communist South and a Communist North. In 1956, South Vietnam, with American backing, refused to hold unification elections. By 1958, Communist-led guerrillas, known as the Viet Cong, had begun to battle the South Vietnamese government.

To support the South's government, the United States sent in 2,000 military advisors--a number that grew to 16,300 in 1963. The military condition deteriorated, and by 1963, South Vietnam had lost the fertile Mekong Delta to the Viet Cong. In 1965, President Lyndon Johnson escalated the war, commencing air strikes on North Vietnam and committing ground forces--which numbered 536,000 in 1968. The 1968 Tet Offensive by the North Vietnamese turned many Americans against the war.

The next president, Richard Nixon, advocated Vietnamization, withdrawing American troops and giving South Vietnam greater responsibility for fighting the war. In 1970, Nixon attempted to slow the flow of North Vietnamese soldiers and supplies into South Vietnam by sending American forces to destroy Communist supply bases in Cambodia. This act violated Cambodian neutrality and provoked antiwar protests on the nation's college campuses.

From 1968 to 1973, efforts were made to end the conflict through diplomacy. In January 1973, an agreement was reached; U.S. forces were withdrawn from Vietnam, and U.S. prisoners of war were released. In April 1975, South Vietnam surrendered to the North, and Vietnam was reunited.

Source: Overview of the Vietnam War
Digital History ID 2925
http://www.digitalhistory.uh.edu/era.cfm?eraid=18

13

Richard Burns
Chief Warrant Officer
United States Army

Vietnam War

Young Warrant Officer Burns

What a night. Dawn was approaching and tracers and explosions filled the air. About 10 o'clock the previous night the mortaring and gunfire began. Whoever was firing was putting on quite a show. It was my turn to fly that morning and I thought about getting sleep but it was useless to try. I was a sector pilot flying the Cessna L19/O1 Birddog, a single engine, two seat observation aircraft. Along with CPT Norm Wood, we would take rotating shifts one flying in the afternoon and then the next morning and then the other flying in the morning and the next afternoon. We lived in a small structure behind the main house near the center of My Tho, the capital city of the province of Dinh Toung. Dinh Toung province was about 20 miles southwest of Saigon. Its southern border was the upper river delta of the Mekong River and to the north was the Plain of Reeds. The province was flat, mostly rice patties crisscrossed with canals. Highway 1, a major route from Saigon to the southern delta, was the main route for equipment and supplies to the South Vietnamese Army in the Delta area. One of our primary duties was to fly over the highway every morning to check for damage caused by the Vietcong the night before, hence the rotating schedule. Other times, we were free to go out on our own to look for the Viet Cong, or support Army of the Republic of Vietnam (ARVN) troops. I wanted to fly all my life, and at age 20, it seems my dream had come true.

Cessna L-19/O-1 Birddog

I was interested in being a pilot since I could remember. I was born and raised on Long Island, New York, near Mitchell Field and Roosevelt Field where Charles Lindbergh started his famous trip to Europe alone in a single engine airplane. Long Island is known as the 'Cradle of Aviation' not only because of Lindbergh, but because many early aviators flew there, including Glenn Curtis and Jimmy Doolittle. Long Island was also a major aviation industrial area where Grumman Aircraft and Republic Aircraft factories built a great many of the aircraft used to win World War ll.

At 16, I started taking flying lessons at Zahn's Airport, located in Amityville, Long Island, in a Piper J3 Cub. I didn't have a driver's license so my mother had to drive me to the airport every other Saturday. That was all I could afford since I was still in high school and working part time in a grocery store. I soloed in November, 1962, and earned my private pilot's license soon after. When I graduated high school, there didn't seem to be a lot of opportunity for me to work as a pilot, so I investigated the armed forces. The Army was the only service that would let you be a pilot without going to college. Seemed right to me and that is the way I went. On August 9th, 1965, I left home for Ft. Dix, New Jersey for basic training. My mom wasn't too thrilled, having been through World War II, with my father, who had passed when I was 16, but she gave me her blessing anyway.

The Army recruiter said I would have a better chance of going to flight school if I was already in the Army, so my "MOS' (Military Occupational Specialty) was helicopter mechanic. After basic training at Ft. Dix, New Jersey, I was off to Ft Rucker, Alabama to learn how to fix helicopters. As soon as I could, I applied for Warrant Officer Candidate Flight School. I had gotten married to my high school sweetheart, Portia Perillo, on December 26, 1965 during Christmas leave and introduced her to Army life and living on a Private's pay check, about $98.00 a month. We lived outside the gate at Ft Rucker and rented a trailer to live in. I went to school every day while she tried to make the best of it after moving from New York to Alabama.

Thankfully, I was accepted into the Warrant Officer Program and in April, 1965, I was sent to Ft Wolters, Texas to begin rotary wing flight training. My new wife moved back to New York to live with her parents. The Vietnam War was building up and there was a great need for pilots, I ended being in a class of about 500 other WOCs (Warrant Officer Candidate). I knew how to fly airplanes but helicopters was a whole new ball game. Then one morning while all 500 of us was standing in formation, the First Sergeant asked for volunteers to go to fixed wing training at Ft. Rucker. I thought that would be easier then learning to fly a new kind of aircraft, so I put my hand up. About another 500 hands also went up. My heart sank when I looked around. Through a series of questions about qualifications, the sergeant asked "Who among you has a pilot's license"? Out of all those people, twenty four of us were chosen to go to fixed wing training, back to Alabama. At Ft. Rucker we were designated as WOFWAC (Warrant Officer Fixed-Wing Aviator Course) Class 66-12. We started ground school and flying in the L19 Birddog, a single engine, two seat observation aircraft. We learned formation flying; did landing and take-offs from small unprepared airstrips; and tactical procedures used by the Army. We were also the first class to receive multiengine and instrument training in the Army T-42, a military version of the Beech Baron. We graduated in February 1967. I was lucky enough to have my new wife Portia pin my wings on me. It was one of the proudest moments in my life.

Back in the Dinh Toung Province, I was now a 21 year old experienced combat pilot supporting the ARVN 7th Division, which was responsible for combat operations in this area. CPT Wood and I supported the ARVN Division and its American advisors, and more often than not carried ARVN army officers (observers) in the back seat, who communicated directly with ARVN troops on the ground. When there were no planned operations, we would fly alone looking for enemy activity. Other U.S. support in this province included Air Force FAC pilot's, and Navy riverboats. The United States 9th Division was also located at Dong Tam, a U.S. base 5 miles west of My Tho. The airport we flew out of was called Binh Duc, about 20 minutes west of My Tho, and along the north side of the Mekong River. Originally it had a dirt runway about 2000 feet long. Later it was covered by steel strips with holes punched through them, called PSP. We had a shed opposite the aircraft revetment, built out of dirt filled ammo boxes, with a make shift latrine behind it, a 'one holer.' There was a table and chairs made out of the same dirt filled boxes.

On January 31, 1968, the Vietnamese were celebrating Tet, their most important holiday of the year. However, this year, the North Vietnamese and Viet Cong (VC) army decided to do major battle. In the Battle of My Tho, their main attack came from the north and west of the city. By chance, it was my turn for the early flight. At first light, I was wondering how my crew chief, Ron Estes, and I were going to get to the airfield to take off. By now, there were small arms and automatic weapons fire in the streets of My Tho. You could hear the bullets buzzing through the trees and hitting the buildings.

A Vietnamese Army armored car, which really looked like a tank with tires, was made available to us. Mounted on top was a 50-caliber machine gun. I didn't have too much information about what was happening, other than from my own observations. The VC were advancing through town, something I had not seen before nor ever expected to see. Ron and I entered the vehicle through a hatch in the top and settled in for a ride to the airport. An ARVN soldier was sitting on the lip of the entranceway of the tank manning the machine gun. Suddenly, the ARVN soldier swung around and started firing. We had driven only for a few minutes and were only down the street from where we started. The noise was deafening. He had spotted a squad of VC walking towards us like they were on a Sunday stroll. The car accelerated as we headed towards the airport.

After about 20 minutes of driving, the ARVN sitting in the hatch suddenly dropped head first into my lap, immediately followed by a loud explosion. As quick as he came down he went back up and began firing his machine gun. I remember holding on for dear life and wondering what was going on. The VC had fired a B40 rocket, narrowly missing us. Finally, we made the turn into our operations hut.

At the airfield, I met my Vietnamese observer, a Lieutenant, who seemed to know no more than I about what was going on. Ron loaded up the rockets and armed the M60 as I did the preflight of the aircraft. The O-1 Birddog aircraft was designed as a single engine light reconnaissance aircraft. But my aircraft was fitted with six 2.75-high explosive rockets. There were four on the left wing and two on the right wing, along with an M60 machine gun. It was a "little" airplane with a big "sting." Looking over the back of the revetment my aircraft was parked in, towards the north side of the runway, I could see ARVN armored vehicles in a line between us and the tree line from which the Vietcong were firing at us. I started up and made a quick turn out of the revetment and took off to the west, making a left turn out over the Mekong River to fly back over the city of My Tho.

Warrant Officer Burns with Cessna L-19/O-1 Bird Dog

Reaching 1,500 feet over the city, the sky turned red with tracers. I did hear the bullets hitting the airplane causing a 'thumping' sound with each hit. There were scattered clouds that I climbed up into to hide. In the cloud, I made an immediate right turn, hoping to avoid enemy fire. My ARVN observer was talking to his counterpart on his radio, as I talked to the American advisors on the front lines facing the VC to the west. When I came out of the clouds, I saw the Vietcong amassed in the western half of the city. This was confirmed by radio communications with the American advisors on the ground. I made several rocket and machine-gun runs against obvious targets. I had a clear view of enemy movement and was in continuous communications with our troops. When I was out of ammo, and low on fuel, I returned to Binh Duc airfield. While refueling and reloading, my Vietnamese observer informed me he was not going to accompany me again. I remember being mad at him and telling him that we were fighting for his country and to get in the airplane, but he still refused. I took off without him.

During that day, I continued observations and reports over the city. Every time I fired my rockets and machine gun, I received heavy small arms and automatic weapons back. Later I counted 13 bullet holes in my aircraft. Thank goodness for self-sealing fuel tanks. As I was low on fuel, and night was settling in, I was told that the Binh Duc airfield was not secure. I flew west to Vinh Long airfield, which was my company's main base, and was informed that that airfield was also not secure. I then flew south to Can Tho and landed. Night had settled in I was told I could rest in a room with an empty bunk, but I could see by family pictures and personal items, that this was someone's room. When I said I didn't want to take this fellow's bunk, I was told that the officer whose room this was would not be returning, ever. I didn't get any sleep that night either.

Citation:

> "The President of the United States takes great pleasure in presenting the Distinguished Flying Cross to Chief Warrant Officer Richard G. Burns for heroism while participating in aerial flight evidenced by voluntary action above and beyond the call of duty. Chief Warrant Officer Burns distinguished himself by valorous actions near My Tho, Republic of Vietnam. Although the road from the city of My Tho to the Ben Duch Airstrip was unsecure, Warrant Officer Burns volunteered to make the trip. He drove to the airstrip in an armored car and managed to get to the field after his vehicle was hit by an enemy rocket. Once there, he lifted off under heavy ground fire. Throughout the day and into the night he provided the only aerial observation available to the city of My Tho. He adjusted artillery fire and on numerous occasions engaged targets with his own armament. His aircraft was constantly under heavy ground fire resulting in several hits. He continued to cover the besieged city leaving only to return to his unsecure airfield for fuel and armament. His actions were in keeping with the highest traditions of the military service and reflect great credit upon himself, his unit, and the United States Army."

I flew 11 hours on the 31st of January in support of the troops on the ground in My Tho. Through the following weeks, long days were routine. By late February, after flying over 80 hours in 3 weeks, things had gotten back to near normal. I remained as a sector pilot until July 2nd, 1968, when I was shot down and wounded supporting an ARVN Intelligence Platoon. I returned to the States and was hospitalized at St. Albans Naval Hospital in Queens, New York.

I was released after 7 months and was able to return to flying duty with the 52nd Artillery Battalion which maintained the Nike and Hercules anti-aircraft missiles around New York City. I flew support missions and VIP flights up and down the east coast. With six months to go on my obligation, I received orders to return to Vietnam. Fortunately for me, I didn't go. The war was winding down and I guess they didn't need a lot of pilots. I left the Army at the end of my obligation in February, 1970.

I tried to find a flying job with the airlines, but with the number of pilots available, a gas shortage which affected anyone trying to find a job, and a family to support, I took a job with the New York City Transit Authority as a Special Investigator. During my seven years there, I kept flying on my off hours, earning my Certified Flight Instructor certificate. I had a part time job teaching at a flight school during the weekends at Long Island Islip Airport. Not caring to work in the subways, I applied for, and was hired as an Air Traffic Controller at New York Center, located on Long Island, New York. I worked there almost 5 years when the controller's union went on strike in 1981. I, along with over 11,000 others, was fired by President Reagan.

I continued giving flight training where ever I could find students. One of my students worked with AIL Systems, a defense company on Long Island, who had a contract with the Air Force on the B-1B Bomber. He recommended me for a position with the company as Repair Manager of the advanced defensive electronics hardware used on the B-1B Bomber. I started working there in 1992. I continued going to school at night and earned and Associate, Graduate, and Master's degree. Working on a steady basis allowed me to open and operate Aviation Services, Inc., a flight school at Republic Airport, in Farmingdale, long Island, N.Y. I worked at AIL for 14 years during the week, and at the flight school at night and weekends. Also, I developed aircraft cockpit utility lighting which received FAA Supplemental Type Certifications (STCs) on 42 different general aviation aircraft.

As the B-1B program slowed down, and after being out of work as an Air Traffic Controller for 14 years, the Clinton Administration allowed us to reapply to the FAA as controllers. Not expecting to be hired back, I learned of a position as a Flight Standards Aviation Safety Inspector in Charlotte, North Carolina. I was hired in 1996, and left AIL Systems, traveling to the Flight Standards District Office in Charlotte. After almost 5 years, I transferred to the New York FAA Regional Office as an Air Carrier SME (Subject Matter Expert). After a year at the Region, I transferred to the New York Flight Standards District Office in Garden City, NY, where I worked as the Principal Operations Inspector (POI) overseeing two worldwide cargo companies operating the largest B747 fleet in the world. During my work as an Inspector, I earned type certificates in the BE3100, Learjet, and B747. I gave type certification checks to over 150 airmen, and travelled around the world. In 2009, because of medical reasons, I retired after 22 years of government service. I moved to The Villages here in Florida that same year.

I belong to several veterans' groups, including the Distinguished Flying Cross Org. I am an FAA Safety Representative, conducting safety seminars to pilots in central Florida, and the Safety Officer in The Villages Aviation Club. I was awarded the FAA's Wright Brothers "Master Pilot" award for over 50 years of safe flying in February, 2014. I continue to fly locally offering free rides to veterans as my way of showing appreciation for their service. I will fly for as long as I can do it safely. It has been too long to give up now.

Mark Laughridge (FAA) with Rich and Portia Burns, 2014

Rich Burns also helped greatly with this book; he provided some original ideas, helped proof-read other author's chapters and was always available to help. Rich continues to teach aviation safety and standardization to young pilots and has a pro bono program to help WW II and Korean War veterans get a chance to fly in central Florida.

14

Jack E. Deal
Lieutenant Colonel
United States Air Force (Ret)

Vietnam

Jack Deal

I was born March 11, 1933 in Muncie, Indiana to Archie and Eva Deal. My life started with a bang and has continued in that manner ever since. The Muncie Press put my Birth Announcement on its front page with the Headline of the Article stating, "New Deal Comes to Muncie". In reference to President Roosevelt's New Deal announcement the previous week.

I attended Burris Laboratory School from 1939-1951 completing Kindergarten thru Graduation from High School and began my adult life. College was next, attending Ball State University, and Graduating in 1955 with a Bachelors' in business. Then, in 1956 I went on Active Duty with the United States Air Force; from that point on literally traveled the World; serving my Country as a member of the Air Force for 20 years, with Tours of Duty in Vietnam and Thailand.

After graduating from the Air Command and Staff College with honors, in the Summer of 1970, my first assignment was to Da Nang Air Base, Republic of Vietnam as a Pilot with the famous 18th Special Operations Squadron to fly the Stinger (AC-119) Gunship.

Excited to the core to be a Pilot with this famous squadron. The Gunship's primary mission in Vietnam was the interdiction of Viet Cong troops and North Vietnamese soldiers along the Ho Chi Minh Trail. Our goal was to interrupt and destroy supplies and troops flowing South. Trucks and Tanks were the primary targets. All missions were flown at night under the cover of darkness (so we thought - laughingly). During every mission we were fired upon by 35- and 57-mm anti-aircraft artillery (AAA) and from Russian made surface-to-air missiles (SAMS).

The AC-119 Gunship fired its weapons while flying a 360-degree turn while also holding a 30-degree bank. The aircraft had two sets of weapons: 1) A bank of six 7.62 mm mini guns, capable of firing 28,000 rounds per minute, and 2) A set of two 20 mm cannons, capable of firing 2,000 rounds per minute.

AC-119 Stinger Gunship over Da Nang, Vietnam

My first combat mission occurred on my first night at Da Nang. We flew "over the fence" (crossing the Vietnam boarder) into Laos and towards the unknown. Within minutes, the sky lit up like nothing we had ever seen. It was B-52's on Arc Light missions (Operation Arc Light was the code name given to the use of B-52 Stratofortress bombers during the Vietnam War). Now I was ready for whatever. Flying in darkness, the gunners held a small flashlight in their mouth so they could see to do their job. The most remembered thing from that mission was the excitement of being there thinking that was a hell of a light show. Realizing later the gravity of the situation when a few, these shells came our way. From that mission on, our crew was known as the "Polish Bandits" and we set records for destroying vehicles and tanks. We alternated between firing the 20 mm cannons and mini-guns from 8,000 feet and below, while evading AAA all this time. Frequently we would break-off and then reacquire the target. The largest convoy we attacked was 31 vehicles and we destroyed all but one.

Our battle damage rules were that a target had to explode or burn to classify it as destroyed. If we shot it and it didn't burn, it was Classified as damaged. Both the video images on the console screens with the audio were recorded on tape. All tapes were sent to Intelligence after each mission.

We always had two F-4 Fighters as escorts on every mission. Their primary mission was to suppress AAA being fired at us. One flew 2,000 feet above us and the other flew at 4,000 feet. We had a funneled light on top of our fuselage, which the F-4 Fighters used to identify and track us. Their secondary mission was to drop laser guided bombs on selected targets.

When the AAA got heavy, our escorts would fly a strafing pattern following one another or one at a time. Our escorts had to keep their eyes on the target from which the AAA tracers were originating. On one occasion, we lost an escort who hit a mountain just below us. No words were spoken but we stayed on the target that night to send a message to those below that we were seeking revenge and would be there until the last shell was fired. We took a few hits that night but we were united and determined to finish the job at hand.

Although I was awarded the DFC on four missions, one that I recall most vividly took place during the monsoon season (when thunderstorms are most frequent and severe) on May 15, 1971. Da Nang was practically under water along with the rest of Vietnam making Viet Cong movement on foot, truck and tank very slow. We are now going after lame ducks (unescorted/unprotected) or what we thought were lame ducks. When we took off from Da Nang that night, little did we know that the enemy lay awake ready to throw anything and everything at us. Flying into the heavy rain was disconcerting. To top it off there were heavy clouds and severe lightning everywhere.

Suddenly we spotted a few hot spots on the infrared radar (a method for detecting and tracking objects which give off heat). It was just what we were looking for and we flew towards the enemy. Flying in and out of clouds trying to get a good shot. Finally, we rolled in on the target and began firing. We set fire to what appeared to be several tanks or large trucks. Suddenly AAA comes up from below en-mass, seeing it coming directly at me, it does not look good. It is 57 mm rounds and seeming going to hit me in the face or so it seemed at that instant. We took a hit 24 inches from me. It rocks the aircraft. We take another hit somewhere near the tail of the aircraft, sparks flying everywhere. Smoke and hydraulic fluid is pouring down the GTPU (Gas Turbine Power Unit), the co-pilots control cables are gone. I look up and see stars where the top of the fuselage used to be. The seat where one of the Gunners sat is gone luckily, he had just moved.

We are in big trouble with a slim chance of survival. Reaching for my parachute, it was no longer there, seemingly blown from the plane. Then reached to unhitch my harness to realize I had already unhitched it. Realizing I had already put on my parachute.

So, Survival Training comes in handy after all, huh?

Now it's time to act! We had taken a 57 mm AAA shell thru the belly and one just thru a wing tip and we are still in the air but badly damaged, mountains closing in and the enemy below. Our F-4 escorts were telling us to bail out because we were pouring fuel all over the sky and we were in grave danger of blowing up at any time. The whole crew decided we did not want to be captured if we were able to survive the parachute landing. We would take our chances. Da Nang AB here we come!

After evaluating our situation, it was decided we had an outside chance of making it home if we didn't blow up or catch on fire in the process. We still had a long way to go and not much chance to get there. As we approach Da Nang AB, we discover that the gear would not come down due to AAA damage. Now we are really in trouble. Trying to land not knowing if we were landing with or without landing gear and not knowing how much fuel we had left remaining. These were problems of major concern. If we crash landed, we would basically close down a major runway at the busiest air base in Southeast Asia.

One of the crewmen had gone below and was able to hand crank the landing gear down but could not lock it in place. Decision was made to give it a try. Nearly everyone holding on in the rear of the plane, where they would have at least a chance if a spark ignited the remaining fuel. We prepared for a wheels-up landing not knowing what the end result would be. We made it down, rolled to a stop, turned on to a taxiway all the while leaking fuel and oil. The crew of 10 departed the aircraft immediately and celebrated. Rocket City (Da Nang) never looked better. Just another day at the office with many more to come as we were soon to find out.

I went on to be awarded four The Distinguished Flying Crosses, the Air Medal with six Oak Clusters, The Meritorious Service Medal, The Army Commendation Medal, Air Force Commendation Medal, The Republic of Vietnam Meritorious Service Medal of Gallantry, National Defense Medal and Expeditionary Medal among others.

Carried many Celebrities including Bob Hope, Jayne Mansfield, Bing Crosby and Roy Rodgers to entertain the troops overseas. Participated in the Eniwetok Atomic Bomb Operations, flew some 175 Combat Missions in Vietnam, Thailand and places that cannot be mentioned. Flew US Diplomats to every corner of the World, flew the NASA Support Missions such as Apollo Series, spent the rest of my career educating and training Officers at the prestigious Air University in 3 assignments encompassing 10 years.

Jack receives service award

Upon leaving Active Duty from the Air Force in 1976, we moved to Montgomery, Alabama. I retired as a Lieutenant Colonel and went into the Real Estate business.

Becoming a Pioneer as one of the first Realtors in the country to command 100% of the Commission when closing a Real Estate transaction, I set records for Sales and Listings in the State of Alabama and become the all-time leader in sales, closing over the billion dollar mark, being featured in a National Television Program about Real Estate Mega Stars, and was named one of the top 200 Realtors in America along the way.

I am not sure I deserve any honors but I have arthritis and cancer caused by Agent Orange and I do not deserve those either.

After my full Retirement in January 2000 I took up Golf again and fortunately found the right outlet, The Villages, Florida: the nation's fastest growing community and a Golf Haven. As of this writing, I have shot my age over 700 times and have set 2 Country Club records.

I attribute any success to my upbringing, Burris School and my father. The New Deal actually came to Muncie. Not sure I deserve this new honor, but Arthritis and Cancer caused by Agent Orange have affected me. Do not deserve either of them.

Jack with his wife in The Villages, FL

15

Frederick (Rick) L. DeLoach
1st Lieutenant
United States Army

Vietnam War

Young Rick DeLoach

I was born in Miami Beach, Florida, December 10, 1947. I grew up around aviation as my father worked for Eastern Air Lines in Miami. I had a love for aircraft all of my life, so when I entered military service, I felt that it was meant to be, that I fly some type of aircraft.

I completed Army Reserve Officers Training Corps (ROTC) while finishing college and graduated with a Bachelor of Science Degree. I then joined the United States Army as a Commissioned Officer and was trained in the Medical Service Corps. Afterwards, I completed Helicopter Flight School from January to October 1970 in two stages. The first stage was Primary Helicopter Flight School at Fort Wolters, Texas, followed by Advanced Helicopter Flight School at Fort Rucker, Alabama. Immediately after graduating, I attended Aircraft Maintenance Officer School in October 1970, at Fort Eustis, Virginia.

On March 22, 1971, I flew out of Fort Dix/McGuire Air Force Base, New Jersey into Bien Hoa Air Base, in South Vietnam, just northeast of Saigon. The aircraft refueled twice en route; first in Anchorage, Alaska and the second in Yokota Air Base, Japan.

I was assigned to the 101st Airborne Division as a Medevac pilot with the 326th Medical Battalion.

After arriving at the 90th Replacement Depot in Long Binh/Bien Hoa, I had to wait several long days for a unit assignment. Because I had been trained both as a Medical Evacuation helicopter pilot and also as an Aircraft Maintenance Officer, it apparently made the assignment process more difficult. After three days though, I was assigned to the 101st Airborne Division as a Medevac pilot with the 326th Medical Battalion, also known as "Eagle Dustoff". 'DUSTOFF' is an acronym for: Dedicated Unhesitating Service To Our Fighting Forces.

Eagle Dustoff

Eagle Dustoff Huey on stand-by

The unit was located at Camp Eagle, which was west of the town of Phu Bai, north of Da Nang, and south of the Imperial Capital of Hue. Camp Eagle was situated less than thirty-five miles south of the Demilitarized Zone (DMZ) between South Vietnam and North Vietnam. Since there were already two aircraft maintenance officers assigned to the unit, I was assigned to fly Medevac missions, due to my training and branch assignment of Medical Service Corps.

I flew many missions from our headquarters at Camp Eagle, but I also flew missions from Camp Evans. Eagle Dustoff kept at least one UH-1H "Huey" helicopter on standby at Camp Evans, which was approximately a fifteen (15) minute flight north of Camp Eagle. This was done to reduce the flight time to potential pickup sites on the northern edge of our operational area, with some pickup sites close to the DMZ.

Every one of the helicopters received some type of battle damage.

Medevac, or Dustoff, signature mission of the Vietnam War, saved thousands of lives.

On the night of April 15th, 1971, three of Eagle Dustoff's helicopters tried to initiate pickups of wounded soldiers at a field location in the mountains, but they were met with very effective enemy sniper fire. Every one of the helicopters received some type of battle damage. Our standby crew listened, intensely, to the radio conversations between our fellow pilots and operations, concerning the current medical extraction attempts. We soon received word from our Company Commander that we were to report to Fire Base Birmingham at 6:00 am the next morning to join two other helicopters from our unit.

I was flying as pilot in 'Dustoff 91,' which was the call-sign of the Aircraft Commander, Richard Dipboye. We were number two in a flight of three helicopters. Once we arrived at the field location, the first helicopter crew started an approach into the pickup zone, but took heavy sniper fire, just like the aircraft had experienced the night before. The planned extraction was aborted temporarily while all three helicopters circled a short distance away from the pickup zone.

A U. S. Air Force B-57 "Canberra" was called in to bomb the area surrounding the wounded soldiers to suppress enemy fire. After the B-57 completed the bombing of the enemy positions, the first helicopter crew made their second approach to the pickup zone. The jungle, that the wounded soldiers were located in, did not allow for the helicopters to land to the ground, so each of our helicopters had to hover above the tops of the shredded trees and perform a hoist operation. Constant artillery explosions had stripped the trees of leaves and twigs from the branches, so all that was visible was the bare trunks and jagged limbs on very tall trees.

We held our position no matter what…

Each of our helicopters carried an electric hoist, mounted on the right side of the cargo area, behind the right pilot's seat. Attached to the hook was a "jungle penetrator," which was used to penetrate through the trees to reach the wounded soldiers. Once the jungle penetrator reached the ground, the three seats were folded down and wounded soldiers were seated on the penetrator to be hoisted back up to the hovering helicopter.

During hoist operations, the helicopter and its crew were extremely vulnerable to enemy fire, because the aircraft cannot move while hoisting wounded troops out of the jungle. Any movement of the helicopter in any direction would drag the wounded soldiers through the limbs and tree trunks, further injuring them, so we held our position no matter what was fired at us.

During each of the hoist operations, there were two AH-1G "Cobra Gunships," constantly firing hundreds of machine gun rounds past either side of the Medevac helicopters. This was partially done because Medevac helicopters were UNARMED and unable to defend themselves when fired upon by the enemy. It was a strange feeling, to both be stressed by the machine gun bullets passing so close on either side of our aircraft, while at the same time, being comforted by the protection that the bullets provided to us.

The first helicopter crew hoisted up two wounded soldiers and departed to drop them off at the closest hospital, usually a Mobile Army Surgical Hospital (MASH). Our helicopter was next in turn and we hoisted up two more wounded soldiers and transported them to the hospital. The third helicopter crew hoisted yet another two wounded soldiers off the smoldering jungle floor. By that time, the first helicopter crew had returned and hoisted another two wounded soldiers to safety. In total, eight wounded soldiers were hoisted to safety that day and transported for treatment at the hospital.

The pilots that flew the three Medevac helicopters were each awarded with the Distinguished Flying Cross Medal (DFC) for the mission, of which I was one.

The next day, still operating out of Camp Evans, I was flying a mission to pick up a wounded soldier that had set off a Booby Trap explosive. Just prior to landing, the rotor-wash of our helicopter's main rotors set off another explosion from a hidden booby trap in the tall grasses. Upon examination of the fuselage of the helicopter, we found that there were shrapnel holes throughout the tail boom and cabin areas of the helicopter. After we transported the wounded soldier to the hospital, the aircraft had to be taken directly to maintenance for evaluation. The extensive damage forced maintenance personnel to swap out the tail boom with one of the less damaged helicopters from the earlier rescue mission. Such dedication by the maintenance personnel kept as many aircraft flying at all times, and in so doing, saved many lives in the Vietnam War.

Citation:

"The President of the United States takes great pleasure in presenting the Distinguished Flying Cross to First Lieutenant Frederick L. DeLoach for heroism while participating in aerial flight evidenced by voluntary action, above and beyond the call of duty in the Republic of Vietnam. First Lieutenant DeLoach distinguished himself on 16 April 1971 while serving as Pilot of an unarmed UH-1E medical evacuation helicopter, Dustoff 91, during the emergency extraction of casualties near Fire Base Bastogne in Thua Thien Province, Republic of Vietnam. On the previous night, three aircraft made numerous attempts to extract the casualties and were severely damaged by enemy fire. Lieutenant DeLoach volunteered for the mission expecting to receive hostile fire and watched as an accompanying aircraft was fired upon. While approaching the extraction site, he also came under heavy hostile fire, but was able to safely extract two patients. His actions were instrumental in saving the lives of the two wounded men. First Lieutenant DeLoach's outstanding flying ability and devotion to duty were in keeping with the highest traditions of the military service and reflect great credit upon himself, his unit, and the United States Army."

Not long after that mission, I was moved into the position of Maintenance Officer and continued in the position until the unit was disbanded and rotated back to the United States in December 1971. I had to transfer all of Eagle Dustoff's twelve (12) helicopters to other remaining units in the 101st Airborne Division. Consequently, I was the last person to leave the unit. It was an uncomfortable feeling, but nothing was going to prevent me from doing my job and getting back home without incident if at all possible.

After returning home, I spent three years at Fort Bragg, North Carolina. I was in the 57th Medical Detachment, flying medical evacuation missions. The flight area consisted of central North Carolina, southern Virginia, and northern South Carolina. The mountainous terrain was not that different from South Vietnam's Northern Highlands as viewed from the air. One welcome difference was that we were never shot at in the latter.

We also supported the 82nd Airborne Division and all of Fort Bragg. I was the unit's Maintenance Officer and then the Operations Officer. After I left the Army, I soon became an Air Traffic Controller at the St Petersburg-Clearwater International Airport and Tampa International Airport, thus staying involved with aviation. My wife and I are both retired and living on the Gulf Coast of Central Florida. Both of our children are married and also live in Florida.

Rick DeLoach Today

16

George (Joe) Duerre
Captain
United States Marine Corps

Vietnam War

A young Joe Duerre

I grew up in Eau Claire, WI. Post-war America in a small town in the Midwest was an almost idyllic life. Great economy, rock & roll music and my '48 Ford. The cloud on the horizon was the Cold War. But we really never gave it much thought. When I started college most of us had never heard of Vietnam. Four years later it was a fact of life for young American men.

When I graduated from college in 1966 the military draft was the law of the land. There were a number of options if a young man did not want to go into the service. He could get a teaching degree, run to Canada or get a job with a defense contractor. I felt that if I did not stand up for my country, I would have a hard time living with myself for the rest of my life. So, I volunteered. I wanted to fly and the Marines gave me a contract that said if I completed Officer Candidate School and was commissioned, I would go to flight school. If I completed flight school, I would serve 4 ½ years on active duty and have a 1 ½ year reserve commitment.

I entered the Marine Corps in August 1966 and after three months at OCS in Quantico, VA, was commissioned a second lieutenant. I then completed flight training at Pensacola, FL and received my wings in January 1968. Because of the rapid escalation of the war in Vietnam and the attrition of Marine helicopter pilots, all new Marine aviators were assigned to helicopter training. My dreams of flying F-4s were dashed. I was ordered to MCAS El Toro, CA for Huey training and then deployed to Vietnam in May, 1968. However, if you had to go to Vietnam and fly helicopters there was nothing better than being a Huey gunship driver.

Joe Duerre and Huey gunship

In Vietnam I was assigned as a Huey gunship pilot in VMO-2, (Marine Observation Squadron) MAG 16, (Marine Air Group) 1st Marine Air Wing. Our squadron was based at Marble Mountain Air Facility just east of Danang.

Marine Corps UH-1E gunship

On April 3, 1969, my assigned mission was flying gunship escort for an H-46 Medevac helicopter. My co-pilot was 1st Lt Alan Nelson. We were on standby waiting for our next Medevac when I got a radio call for emergency assistance in support of the defense and subsequent extraction of a nine-man Marine long-range reconnaissance team (small teams of Marines who had been dropped behind enemy lines to conduct surveillance and relay information on enemy troop movements). What had been the routine insertion of the recon team had turned to disaster. As the team had exited the transport helicopter in the landing zone, they had come under intense automatic weapons and grenade fire by a large North Vietnamese Army (NVA) force.

I left my Medevac H-46 to wait for another gunship and then joined two Huey guns already on station over the beleaguered Marines. We began a sustained and coordinated attack on the NVA that eased the pressure on our troops and allowed them to move about 50 meters to a better landing zone for pick-up.

Gun and Rocket Run

Because of the proximity of the NVA to the Marines, we were executing extremely low-level machine gun and rocket runs. At this point the Hueys were literally flying in the treetops in the face of intense enemy fire directed at us. When the pressure on the Marines eased, we called an H-46 to lift the recon team out. As I had white phosphorous smoke rockets in one of my rocket pods, I flew at a very low level between the Landing Zone (LZ) and the NVA and laid down a smoke screen to cover the H-46 on his approach and landing.

With the smoke screen in place, the Hueys formed up on the 46 and flew alongside of him as he made his approach into the landing zone. Each Huey gunship had four forward-firing .30 caliber machine guns and two movable .30 caliber machine guns fired by the door gunners. As we escorted the 46 into the zone, all guns were firing … a total of 18. Because of our earlier coverage of the recon team, we had a pretty good idea where the NVA were located and that was where our fire was directed. It would have been a foolish NVA soldier to attempt to fire on the 46 at that point. The tactic was effective and we pulled the Marines out of a really nasty situation without a single loss.

The other two gunship pilots and I and the H-46 pilot were each awarded the Distinguished Flying Cross for our actions that day.

Later that month, on April 23, I was the flight leader of two Huey gunships and we were flying gun cover for H-46s doing routine re-supply for the 7th Marines about 20 miles south west of Dining. My co-pilot was 1st Lt. D.J. Simmons and my wingman was 1st Lt. Ken Pennington with co-pilot 1st Lt. Ken Johnson. Pennington was KIA flying a Huey Cobra on Sept 24 '69.

We were contacted by air control and reassigned to the mission of providing cover for the emergency extraction of an eight-man Marine long-range reconnaissance team that was surrounded and heavily engaged in combat with an estimated NVA platoon. Our Marines were outnumbered almost five to one. The recon team had sustained three casualties and one of the casualties had fallen over a steep cliff after having been struck by hostile fire.

As the lead gunship pilot, I was the airborne commander in charge of the rescue mission. When we arrived over the Marines' position our initial task was to provide cover for the team as they moved to recover the man over the cliff. Our two gunships made repeated gun runs above the team at treetop level. All the while we were taking sporadic automatic weapons fire from the NVA. In an encounter like this the door gunners were invaluable. They were our eyes on the sides of the aircraft and were able to direct intense and accurate fire when needed.

When the wounded Marine was recovered, my wingman and I began repeated low level gun and rocket runs to allow the team to move to a more suitable pick-up site. Because there was no cleared landing zone in close proximity, the rescue was to be done by an H-46 equipped with an extraction ladder. The term ladder is used loosely as the device consisted of four 100 foot long chains connected to each other by steel bars placed every two feet or so between the chains. This ladder could be lowered through the treetops as the H-46 hovered above and the team could grab the steel bars and be pulled out of the jungle to safety.

Once again, we used white phosphorous rockets to place a smoke screen between the NVA and the Marines. Because the opposing forces were so close to each other, the smoke rockets had to be placed with precision. This involved a great deal of exposure to intense enemy fire.

As the H-46 maneuvered into place and lowered the extraction ladder, the gunships commenced low level rocket and machine gun attacks against the NVA forces to deter them from firing on the rescue helicopter or the Marines who were being lifted out of the jungle.

CH-46 Sea Knight recovering men via extraction ladder

Everything worked as planned. The transport helicopter was able to lower the extraction ladder and recover all eight men.

For our actions that day, the two Huey gunship pilots were each awarded the Distinguished Flying Cross.

After flying 680 missions and surviving 13 months of combat, I left Vietnam on July 4, 1969. My next duty station was Marine Corps Air Station, Yuma, AZ. I was assigned as one of the station search and rescue pilots which was about as great a job as I could imagine. I worked as an SAR pilot for a 24-hour shift and had 72 hours off. During my work shift I could fly as much as I wanted and took every opportunity to do so. There were four other Lieutenants at Yuma who had flown with me in Vietnam. We were reserve officers with no interest in making a career in the Marine Corps so we were not very serious junior officers. As a result, we had two great years of flying, riding dirt bikes in the desert and being generally grateful for coming home alive. Yuma was a great place to decompress. It was good to have several years to prepare for civilian life in an America that was unsympathetic to Vietnam Veterans.

Because of the rapid escalation of the war in Vietnam and the attrition of Marine helicopter pilots, all new Marine aviators were assigned to helicopter training.

I was discharged in Dec. 1971 and spent the next ten years in marketing with a three-year stint at owning an advertising agency. I sold the agency in 1982 and began a career in Health Care Management. I stayed in health care until retirement in 2001. I became recognized as an innovator in the development and operation of HMOs and became CEO of two different organizations. My years in the industry were both satisfying and rewarding.

Joe Duerre in retirement

I have always felt that a great deal of my success in whatever venture I tried, professional or personal, could be attributed to my training and experience in the Marine Corps.

17

Joseph R. Finch
Lieutenant Colonel
United States Army (Ret)

Vietnam War

A young Captain Joe Finch; second Platoon leader

My father, an estimating engineer, had various jobs in the US, Morocco, Spain, Greece, Italy, Indonesia, Korea, Thailand and Hawaii. Our family moved around the world to stay with him. We were fluent in several languages. When I was old enough to attend University, I moved to my Aunt Virginia's house in

Tucson, AZ, and completed 11th and 12th grade in one year at Amphitheater High School, Tucson.

Captain Joe Finch with a UH-1 'Huey' Nighthawk

As a high school student in Tucson, in 1962, I joined the Civil Air Patrol. I was able to get some stick time on a few aircraft before I got a license, I decided I liked flying, went through Air Force Reserve Officers Training Corps (ROTC) at the University of Arizona. I didn't quite finish my engineering degree, and was out of school on probation 1966. I got a draft notice while I was in Bangkok, Thailand visiting my parents and family for the summer. Drafted into the Army, Sept 1966, I did well in all the basic skills and training, was sent to Leadership Preparation School, and was accepted into Infantry Officer Candidate School at Ft. Benning, GA graduating as a 2nd lieutenant of Infantry 21 Sept 1967. I saw a chance to get into flying and applied for flight school. My orders arrived in a couple of weeks: "You are ordered to report to 25th Infantry Division in Vietnam; You are permitted Temporary Duty to attend US Army Aviation training at Ft. Wolters TX and Ft. Rucker, AL. en route to your duty station in Vietnam".

I was an average student in flight school, had some trouble with instrument flight training, but I liked the Hueys, and enjoyed flight school.

I arrived in Cu Chi Vietnam on 21 January 1969, just before Tet '69 – the Vietnamese Army's all-out assault on American forces. The CO was happy to see me, had been waiting for a new commissioned officer to assign to the second platoon. My first task was to sign for the second platoon, all the airfield lighting systems, the glide-slope system (for landing in bad weather), and nine UH-1 (Huey) aircraft-almost 7 million dollars' worth of stuff!! I spent several days looking for everything, checking serial numbers and signing paperwork. In the evenings I was hearing other pilots talking about their day and I was ready to start flying in combat. Flight school prepares you to steer and manipulate controls of an aircraft, not how to fight with the helicopter as your weapons system. I was itching to get into the fray.

I had not been through FNG training (new arrivals training), was not yet flying operational missions, still learning the Standard Operating Procedures (SOPs), rules, and three weeks later I had a heck of a time adjusting to the fact the bad guys blew up our base camp at 04:00 on 26 Feb How rude!; I ran out to the Bear Pit (where all our aircraft were parked in protective revetments), and cranked up a Huey to get the heck out of Cu Chi. What I couldn't know at that time was that my early "baptism by fire" created a mindset that lasted for the rest of my tour. North Vietnamese Army (NVA) sappers blew up nine of our Chinook helicopters and all our Petroleum Oil and Lubricants (POL). The bad guys really hated our aircraft and helicopters were at the top of the list. Every morning when I woke up to fly, I knew there was a probability I would be shot at or shot down. (At that point the Vietnamese had a price on pilot's heads of $10,000, which was a lot of money to those guys and they didn't even need the whole body—just the head!)

After that first flight in the middle of the night and my baptism by fire, the CO realized I was ready to start flying, and I did OK under pressure, I got a checkout on the local area with our flight standardization office Chief Warrant Officer Ed Behney, then began flying operational missions as a copilot or "peter pilot". Over the next two months I was shot at a lot and our aircraft was hit several times so I knew what that was like too. Irritating, really.

It is not that I was inured to being shot at, it's just I understood that's what life was like in the war zone. There were days that I was terrified and others where I was just aggravated. Though not rated as an aircraft commander, I was a fairly well seasoned aviator by the end of April, and had received hostile fire to my aircraft many times.

1st Lieutenant Joe Finch

A UH-1 Huey on the tarmac

Most days we flew for long hours dropping off supplies, ammunition, food, water, defoliation in areas of heavy resistance, mail, providing suppressing fire to infantry troops under enemy fire, bringing cold drinks or conducting medical evacuations, whatever the infantry needed. Often, we were being shot at as we went in to resupply, and normally, usually, generally most always, we were shot at when we tried to pick up the wounded. (The North Vietnamese and Viet Cong were not stupid. They knew if they shot down an aircraft the Army would do everything in their power to get the crew and the wounded out. 'Charlie would wait for the next aircraft and try to shoot that one down too.) Often what our guys on the ground wanted was more firepower. They knew we had a different perspective when looking down on the battle, and the experienced infantry gunners who were gunners for our Little Bear aircraft were pretty darn good at figuring out where the bad guys were on the ground and hitting the target.

None of us really knows how many days we flew our butts off trying to support the infantry, flying as many as 27 hours in a row, flying directly into the face of enemy fire because we knew that what we did was crucial to the guys on the ground. It's what helicopter pilots did. It was one of the facts of life in Vietnam.

Normally, there were very few people to notice what we had done and many of those who did notice were too busy to write up the event, so most days we flew what might qualify for a DFC or higher award and the only recognition we got was the appreciation of the guys on the ground. That was good enough for us.

7 September 1969 started like many other days at zero five thirty with the roar of dozens of rotor blades as our Hueys took to the sky. Early morning flights of 18 to 30 aircraft were common. Many times these were the combat assault missions, which took the infantry to enemy locations decided by the G2 (Intelligence) during the previous night. Invariably combat insertions went into known hotspots, which led to firefights early in the day.

That day I was flying with the second platoon as a pilot with a seasoned aircraft commander—Chief Warrant Officer, John Driscoll who was also well experienced with enemy fire. At about 09:00 we responded to a call for help from an infantry battalion (2nd Battalion, 22 Infantry—the famed "Triple Deuce"). The unit was pinned down by enemy fire about 45 km north of Cu Chi. By the time we got on station it was a beehive of activity with an Air Force OV-10 Bronco Forward Air Controller (FAC) trying to manage the action. Our first action was to fly into the middle of all that activity and try to identify where the bad guys were.

We began with a series of low-level passes which might encourage the enemy to divulge their positions – mainly by shooting at *US* so everyone else could see the location of their machine guns. The enemy appeared to be well protected in makeshift bunkers, from which they were able to place withering fire on the US troops. (At times like that we realized just how tough the Infantry Battalion commander's job was. They were fighting to keep their guys alive, kill the enemy, and manage all the helicopter and USAF air support and let everyone know what was going on.)

USAF OV-10 Bronco fires spotting rocket

I believe we made three passes trying to determine the enemy locations along a short stream and bamboo line. That was sort of hairy because the bad guys shot at us… a lot. That didn't bode well for follow-on low-level flights—eventually we were going to get hit. Our door gunner and the crew chief fired at the targets they could see and their orange tracers gave the "friendlies" a good fix on the bad guys. Our machinegun fire at the nests of Viet Cong (VC) was effective and showed the Infantry and the FAC where the bad guys were hiding. The VC seemed to decide it was not a good idea to shoot at us anymore.

Next, we had to extract three wounded US troops. On our way in, there was a proverbial 'hail of gunfire'. It was serious enough that the FAC (often an Air Force Captain) called for suppressing fire to help us stay alive. We called for smoke from the "friendlies" to mark their positions on the ground, and to show us where to pick up the wounded. We identified the smoke by color, (yellow), and set up our approach. At that point we were getting wide-eyed at all bullets flying every which way and the aircraft commander was flying his best telling the Air Force controller which direction we were going—a landing from the east to the west along a small wood line. As the aircraft commander gave me assignments or missions, I would call the appropriate people and let them know what was going on and advised the hospital that we would be bringing in wounded. Typically, we were on three or four radios plus the automatic direction finder (ADF) that was kept tuned to AFN Saigon.

John and I had both been under fire before, but he was the AC and briefed us on what to expect, then focused on the FAC and the ground unit commander's radios. He was busy flying the aircraft and keeping the FAC and the ground unit informed. The closer we got to the ground the more the air seemed to be full of bullets with few places to fly and nowhere to fly safely. We came in hot by that I mean fast (probably at 50 - 60 knots) and flared to a halt with dust flying everywhere. The infantry guys came over with their three wounded troops and we were off in seconds. We were hit a few times while we were stopped, but none of us were wounded. We checked to be sure the wounded were secure, then reported to the FAC we were taking off to the west and would break left at 500 feet.

We couldn't really see it as we broke left, I was looking at the sky; we were both busy but there were lots of bullets flying around down there trying to protect us as we took off.

The aircraft Commander gave me the controls, "You have the aircraft! Get us to the Medevac pad at Cu Chi ASAP!" I put the nose over, went to 100 knots as fast as I could. The aircraft commander spent a few moments thanking people for their support and making sure that everybody knew which direction we were going so we would not get hit by friendly fire. We knew that we were lucky to get out of that alive and were sort of breathing a sigh of relief, but we could hear the firefight intensifying and we knew we'd have to go back.

I made a straight-in approach to the Medevac pad at the 12th Evac Hospital. Medical personnel were on hand to take our wounded and within a few seconds they were on their way into the hospital. While I was paying attention to that, Driscoll was on the radios assessing the situation on the ground. He advised Little Bear Operations of the scale of the conflict so they could get another crew ready just in case we could not handle the situation with one aircraft. We headed over to the re-arm point for more ammunition then headed back into the fray. (Remember that sinking feeling as a kid when you knew you were going to get spanked?)

The firefight had gone nasty, there were more wounded, they needed more ammunition and they needed more fire support. The aircraft commander told me to head back to the same site, I advised Cu Chi tower of our departure, and we headed back into the mêlée. This time we had a better feel for the area and got there in minutes. The Forward Air Controller (FAC often an Air Force Capt.) made a glib comment welcoming us back to the "party." There were five more guys seriously wounded and they put a priority on the medical evacuation so we could see ourselves going back into that same mess again. At that point I believe the pucker factor went up.

We knew in advance what we were getting into, and had escaped injury a few minutes earlier only to try our luck again. That time it seemed they were not firing quite so many bullets at us, and the FAC had some F-4 Phantoms on station to drop napalm on places where bullets emanated from the enemy. We lined up on the same landing site but we came in slightly to the left of our previous path, again came in hot, landed and picked up five wounded and got out of there again in seconds. I flew the return path to Cu Chi, we dropped off the wounded to the hospital, and were getting ready to breathe a sigh of relief. But we were called again. We re-armed, re-fueled and went back to that same miserable piece of ground 5 times, the aircraft took a few minor hits but none of us were wounded, we had flown right down the enemy's gun sights so many times it's amazing we survived. Hueys have no armor except for the "Chicken Plates" on the left and right side of the cockpit to protect the pilots, the pilot's bullet proof vests and a few flak vests scrounged up by the crew which they sat on. Our crew chief (Specialist Marvin Epstein I think?) and the gunner (Specialist 4 Hunter, I believe) were credited with suppressing enemy fire with their M-60 machineguns and were credited with several enemy kills. On one of our runs into that firefight, there were Huey Cobras (AKA 'Snakes') on station to provide close air support for us. Those guys really suppress fire well. With the shark's teeth painted onto the sides and front of the fuselage they looked formidable and, as they were only 48 inches wide, they were hard to hit. Cobras were heavily armed with General Electric miniguns—electrically fired 6 barreled machineguns firing 4,000 rounds per minute—plus they had 24-48 rockets! The Cobras could put a lot of ammo on the ground quick if anyone dared to shoot at them.

The Battalion commander of the unit in contact put us in for the DFC for repeatedly flying into enemy gunfire, suppressing the enemy and saving a total of 25 troops that day. Our crew of four went to bed tired that night. At the end of that day my flight gloves were wet with sweat from nerves, fear and 9 hours of hard flying. We felt good knowing we had made a difference, but to us it was still just another day flying support for the infantry. Our company motto was "Lele Makou No Na Puali"—Hawaiian for "We fly for the troops". We had done that. We had no idea we were put in for an award until it arrived in the company area a few days later. Two days later I was certified as an Aircraft Commander. I had earned my AC status.

DEPARTMENT OF THE ARMY
Headquarters, 25th Infantry Division
APO San Francisco 96225

GENERAL ORDERS
NUMBER 11935 — 24 September 1969

AWARD OF THE DISTINGUISHED FLYING CROSS

1. TC 320. The following AWARD is announced.

FINCH, JOSEPH R. 527-78-7339 1LT INF USA
Co A, 25th Avn Bn, 25th Inf Div
Awarded: Distinguished Flying Cross
Date action: 7 September 1969
Theater: Republic of Vietnam
Reason: For heroism while participating in aerial flight: First Lieutenant Finch distinguished himself by heroic actions on 7 September 1969, while serving as a pilot with Company A, 25th Aviation Battalion in the Republic of Vietnam. While on a combat operation, elements of the 2d Battalion, 22d Infantry came in contact with a large enemy force. Immediately, Lieutenant Finch flew to the contact area to provide aerial support for the infantry unit. Upon arriving on station, Lieutenant Finch began a series of low level passes to determine the location of the hostile emplacements. Having identified the enemy positions, Lieutenant Finch, with complete disregard for his own safety, exposed himself to the hostile fire as he placed devastating fire on the aggressors. After expending his supply of ordnance and returning to Cu Chi to rearm his helicopter, Lieutenant Finch continued his aerial assault on the hostile bunkers. His valorous actions contributed immeasurably to the success of the mission and the defeat of the hostile force. Lieutenant Finch's bravery and devotion to duty are in keeping with the highest traditions of the military service and reflect great credit upon himself, his unit, the 25th Infantry Division, and the United States Army.
Authority: By direction of the President, under the provisions of the Act of Congress, approved 2 July 1926, AR 672-5-1, and USARV Reg. 672-1.

FOR THE COMMANDER:

OFFICIAL:
S. R. Janis
S. R. JANIS
CPT, AGC
Asst AG

JAMES M. CONNELL
Colonel, GS
Chief of Staff

DFC Awarded 7 September 1969

A few statistics are in order:

- Total helicopter pilots killed in the Vietnam War was 2,202. Total non-pilot crew members killed was 2,704
- 2,709 people were killed while in Hueys: 1,074 pilots 1,103 crew members 532 American passengers (we do not have a good count on Vietnamese, their numbers probably exceed the number of American passengers.)
- If you look at the number of pilots who served in Vietnam, we lost about 10% of them. For the number of missions we flew, and the number of times people shot at us it is a miracle the statistics weren't higher.
- Army UH-1's totaled 7,531,955 flight hours in the Vietnam War between October 1966 and the end of 1975. It is my opinion that the Huey, including the Huey Cobra (AH-1G), which had 1,038,969 flight hours in Vietnam, have more combat flight time than any other aircraft in the history of warfare.

(Statistics assembled by Gary Rousch, Vietnam Helicopter Pilots Association VHPA)

LTC Joe Finch, US Army (Ret)

In my one year in Vietnam, I medevacked a total of 826 wounded Troops. The real joy is running into one of them in real life and they remembering me.

I went on to earn 35 awards of the Air Medal and two Silver Stars for days similar to the one described above in Vietnam, then transferred to Thailand, where I flew missions in support of Air America for 4 months. In August 1970 I took a detachment of three Huey's to Hua Hin, Thailand, to rescue survivors of a Typhoon that washed the villagers, their livestock and pets into the gulf of Thailand. We rescued just over 2,000 people (plus a few pet pigs, dogs, chickens, etc.) in three days and nights of flying till we were too tired to safely operate our aircraft, sleeping on a mat for a few hours then off to do it again.

In January 1971, I rotated back to the United States to be a unit commander, completed my degree, served multiple assignments in command and staff positions in the US and in Germany where I met my wife, Monika; I had a great military career, and completed two Masters Degrees.

In 1986 I was assigned to Washington DC as the Inspector General of the Military District of Washington; in 1988 I became operations officer for the Inauguration of George Bush senior—the most challenging position I ever held and more stressful than being shot at in a war zone. For a year I worked on that inauguration - 18-hour days, often with one 22-hour day per week. It took a terrible toll on my body. I completed my flight physical with flying colors on 21 October '89, all results indicated I was in perfect health. On 23 October 1989 I fell, hit my head three times on the way to the floor and suffered a near fatal stroke. Unable to stand, walk, or speak, I had lost my short-term memory, math skills, and higher-level functions...

Almost as exciting as the day I earned my Distinguished Flying Cross is my recovery following that near fatal stroke. As they wheeled me into Walter Reed on Halloween, I was helpless. The doctors took an MRI, were unable to determine the cause of such a massive stroke in a relatively healthy young man (45 yrs. old). After 6 months of physical and occupational therapy, it was clear I could no longer perform as a Lieutenant Colonel. I was discharged at 100% disability and was told my life expectancy was about 18 months. They did not know what caused my stroke but were convinced the next one would kill me. I went through another 6 months of occupational and physical therapy. Disability pay was not enough. We used up all our savings while I was not working. Monika was working three jobs to keep food on the table; but it wasn't enough. In 1990 we were broke. That was 27 years ago.

From that dismal prognosis and bleak financial situation, I joined the Centreville Baptist Church, had 12 months of "choir therapy" and with God's help have made *a remarkable recovery*. I began a new career as a civilian engineer working full time with various engineering firms while completing a 3rd Master's Degree; I was volunteering during my off- time with the wounded troops for 15 years.

In 2009 the US Marine Corps submitted me for the "Spirit of Hope" award for 15,000 volunteer hours in support of troops in the Middle East and the wounded in the hospitals near Washington DC. (The Spirit of Hope award is presented to people who best typify the spirit of Bob Hope in raising the morale of soldiers, sailors, airmen, and Marines). At the same time, my professional life earned 13 major National awards for the programs I was working on. My life has made a difference. In 2014 Monika and I were able to move to The Villages, Florida.

The 2009 Spirit of Hope Award

There is no other term that fits, I have been *blessed* by a loving God.

"But those who hope in the LORD will renew their strength. They will soar on wings like eagles; they will run and not grow weary, they will walk and not be faint." (Isaiah 40:31 NIV).

Joe and Monika Finch enjoy a fulfilling life

"I can do all things through Christ which strengtheneth me." (Philippians 4:13)

Agent Orange

One of the missions we flew as helicopter pilots was to drop a liquid defoliant called "Agent Orange."

None of us had any idea of the devastating effect that would have on every soldier that went to Vietnam.

18

Michael A. Gulick
Major
United States Army

Vietnam War, Desert Storm

Chief Warrant Officer Michael Gulick with DFC awarded 4 years after the fact by a thankful ground unit commander who did NOT forget Michael's courage under fire.

I had been surrounded by the Vietnam War for all of my teenage life and most notably in my high school years. Every night, details of battles, destruction, and the tolls of the dead and wounded occurring thousands of miles away were simulcasted as part of the regular news segment. The problem was that it happened so often, it did not seem real and it was easy to become "numb" to the violence. The war looked dangerous, but at the same time I was intrigued. It offered something most teenaged boys want —adventure and the opportunity to be a part of something greater.

High school was my last vestige of normal life for the next three decades. In January 1970, during my senior year of High School the Presidential Lottery Draft Date Selection took place. The lottery involved a bowl full of balls with birthdays written on them. The first 124 birthdates drawn from the bowl were guaranteed to be drafted, birthdates 125–240 were a possibility and birthdates 241–365 was unlikely to be drafted. The only lottery I ever won was when my birthday, January 31, was the third ball drawn. As I was in the bracket that was assured to be drafted, I opted to enlist in the Army, rather than risk of being thrust into another branch of the Armed Services. This event set my military career in motion.

Once enlisting in the Army, I was given 30 days after graduating High School to report to the recruiting station in Oakland, California by 10 July 1970. The recruiting station processed everyone very efficiently. The Army had medical and other related paperwork administration down to a science and it took only four hours for me to be cleared for duty and placed on a bus heading to North Fort, Fort Polk, Louisiana for Basic Combat Training (BCT).

I finished in the top ten percent of the BCT class and was promoted to Private 2nd Class (E-2). I then started Advanced Combat Training (ACT), which included taking various qualification tests to confirm your capabilities. I qualified for Warrant Officer flight training and this became my new career path. Shortly after, I was promoted to Specialist 5 (E-5) and shipped off to Fort Wolters, Mineral Wells, Texas to start Basic Flight Training. From here the Army transferred me to Fort Rucker, Alabama for Advanced Flight Training, where I graduated 3rd in my class at Flight School.

At this point I was discharged from the Army as an Enlisted Solider and subsequently appointed as a Warrant Officer (WO-1). As a Warrant Officer, my flight specialization continued with a focus on advanced helicopter training in Cargo Helicopters - the CH-47 "Chinook" Helicopter.

CH-47 doing water landings

One of the greatest things about a CH-47 is that they are able to float, and water landings were a required part of our training. Playing boat with a helicopter can be fun…

The Army gave me two weeks leave before I needed to report to Travis Air Force Base, California for deployment to Vietnam as a replacement pilot to support the pilot shortage. That was the last visit I had with my family before embarking on my adventure into the unknown.

In July 1970, I landed in Saigon, Vietnam where I was assigned to Camp Holloway, Pleiku, Vietnam. Camp Holloway was in the Army Second (II) Corps in the Central Highlands, at the intersection of Cambodia, Laos, and Vietnam. Camp Holloway become my base of operations for the next 12 months.

Pleiku, Central Highlands where Cambodia, Laos & Vietnam intersect

Camp Holloway, Pleiku in the Second (II) Corps in the Central Highlands
Main runway made from PSP (Perforated Sheetmetal Planking)

My initial assignment in Vietnam was as a CH-47 (Chinook) Pilot to the 179th Assault Support Helicopter - Call Sign: "Shrimp Boat". I only stayed with that unit 1 week as co-pilot.

The CH-47 Helicopter parting spots made from 55-gallon fuel barrels
Camp Holloway, Pleiku, Vietnam

For the next eight weeks I was assigned to A Company Assault Helicopter/227 Air Cavalry Battalion, 101st Airborne Division, Call Sign: "Chicken Man" as a Utility Helicopter - UH-1H (Iroquois nick named "Huey") co-pilot where I flew regular missions for the Combat Assault Helicopter Company, such as combat assaults (multiple aircraft delivery of combat troops), re-supplying food and ammo to fire bases and ground troops, and "Ash and Trash" (general transport of troops in and out of the mountains).

Warrant Officer Michael Gulick shown with his helicopter waiting for the next assignment

Having completed many successful missions and, most importantly surviving, I was selected to command my own aircraft. I was subsequently reassigned to the 57th Assault Helicopter Company as an Aircraft Commander – Call Sign: "Gladiator 39". I remained an Aircraft Commander for the remainder of my tour in Vietnam.

We were one of the first helicopter units to receive the helicopter mounted TOW missile launch system (which is a tube-launched, optically-tracked, wire-commanded data link guided missile). Using this TOW system, we were able to capture a Soviet T42 Tank during an attack on the city of Kon Tum, Vietnam.

Warrant Officer Michael Gulick next to a captured Soviet T42 Tank

I was also one of the first pilots to be shot at with the newly deployed Soviet SA-7 Grail missile (a shoulder-fired, low-altitude, heat seeking, surface-to-air missile) – a missile that lead to the modification of all helicopters to include engine exhaust diffusers. This modification was nick named the "toilet bowl" and its goal was to reduce the heat signature the engine exhaust caused on the tail boom. By deflecting the exhaust into the rotor system, it removed the hot spot on the helicopter, disrupting the SA-7 guidance and vastly improved the odds of survival.

Engine exhaust diffusers

Wire Strike Protection System

The Army also modified the in-theater helicopters by installing wire cutting blades to the *Roof and Bottom* front of the helicopter. This was done to counteract the enemy's tactic of stringing cables between trees to knock helicopters down in flight. This was significant, as when maneuvering through valleys to avoid detection we normally flew below the tree line. Hitting one of these cables would sever the rotor system controls, causing the aircraft to lose control and often crash killing everyone on board.

Throughout 1971-72, I continued to fly missions all over the Central Highlands. Given the circumstances of my situation, I had little interaction with the people of Vietnam. However, I did have an opportunity to connect with a native mountain group of people called the "Mountain Yards" as part of my job delivering supplies to the mountain region. Several times we were invited to participate in their festivities. It was my perception that they didn't have strong feelings for or against the war, as long as it didn't negatively impact them.

One of our key missions was supporting the American Military Advisors to the Army of the Republic of Vietnam (ARVN) units. In June 1972, my helicopter was assigned to directly support the Senior Advisor, Major Bush of the 23 ARVN Infantry Division in city of Kontum, Kontum Providence, Vietnam for troop support, re-supply and emergency assistance, as needed. On 20 June 1972, in the Western part of the Kontum providence, the North Vietnamese started attacking one of our fire bases manned by the 23 ARVN battalion. As was the norm for this action, my helicopter was placed on high alert, fueled, and sitting with the engine started waiting for the call to action. We would refuel as needed with the engine still running to ensure we had the maximum fuel available and were ready to react, if needed.

The call eventually came. Reports came in that we were taking causalities and the commander asked if I could fly my helicopter, a UH-1H, into the fire base and rescue the wounded, so they could be taken to the medical facilities for treatment. Luckily, the weather that day was good for us – cloudy and the potential for rain. We took off immediately and flew towards the fire base. Other support/attack helicopters, AH-1H Cobra gunships, were already on site to provide support. We circled the fire base assessing the possibility of making a safe extraction from a "hot" landing zone.

The situation at the fire base was not improving and the number of seriously wounded soldiers was increasing, so I made the decision to go in. I took the aircraft out of sight of the fire base, descended down to 20 feet above the height of the jungle and started my approach. I knew that at that altitude the enemy would struggle to determine the direction I was approaching from, so our chances were good of reaching the base before being detected. The escape, however, would be the more dangerous maneuver.

We broke free of the jungle doing 130 MPH and had only 100 feet to stop the helicopter before we overran the fire base. We successfully completed the deacceleration turn and landed. The crew and ground personnel quickly loaded the wounded soldiers acutely in need of medical care, while the enemy fire continued. This must have been my lucky day, as somehow the enemy fire did not hit anything critical on the helicopter. I pushed the engine to maximum power and started our takeoff run. Staying no higher than 20 feet off the ground we soon reached the jungle, which provided some cover from enemy fire. After about two miles I climbed the helicopter to the normal 1,500 feet and headed for the medical facility. In the end, all seven soldiers we rescued and survived – we were credited with saving their lives.

The attack was starting to turn in our favor, but it was not over yet. The fire base was running low on ammunition and medical supplies and required assistance. We flew another mission carrying needed medical provisions and ammunition to the surrounded fire base, helping them to withstand the next attack. I feel fortunate that my crew, the helicopter, and I were able to provide this needed support and make it out unharmed.

Typical Base with the occasional helicopter that did not make it

By chance, four years later while eating at the Officer's Club I ran into Major Bush, now a Colonel, who I supported that day. He asked if I ever received the award he submitted on my behalf as Battalion Commander. Surprised, I told him I had no idea an award was submitted.

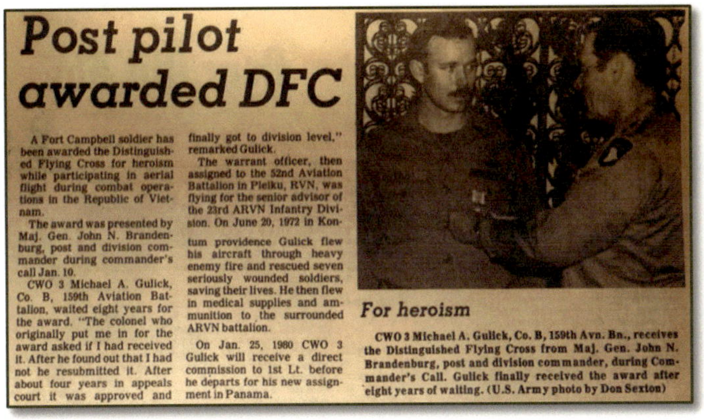

Post pilot awarded DFC

A Fort Campbell soldier has been awarded the Distinguished Flying Cross for heroism while participating in aerial flight during combat operations in the Republic of Vietnam.

The award was presented by Maj. Gen. John N. Brandenburg, post and division commander during commander's call Jan. 10.

CWO 3 Michael A. Gulick, Co. B, 159th Aviation Battalion, waited eight years for the award. "The colonel who originally put me in for the award asked if I had received it. After he found out that I had not he resubmitted it. After about four years in appeals court it was approved and finally got to division level," remarked Gulick.

The warrant officer, then assigned to the 52nd Aviation Battalion in Pleiku, RVN, was flying for the senior advisor of the 23rd ARVN Infantry Division. On June 20, 1972 in Kontum providence Gulick flew his aircraft through heavy enemy fire and rescued seven seriously wounded soldiers, saving their lives. He then flew in medical supplies and ammunition to the surrounded ARVN battalion.

On Jan. 25, 1980 CWO 3 Gulick will receive a direct commission to 1st Lt. before he departs for his new assignment in Panama.

For heroism

CWO 3 Michael A. Gulick, Co. B, 159th Avn. Bn., receives the Distinguished Flying Cross from Maj. Gen. John N. Brandenburg, post and division commander, during Commander's Call. Gulick finally received the award after eight years of waiting. (U.S. Army photo by Don Sexton)

Colonel Bush explained he had submitted me for the Distinguished Flying Cross (DFC) for my efforts that day at the fire base and he would see to it that these actions were not lost. In 1979, four years after our meeting, Major General John N. Brandenburg, post and division commander, 101st Airborne Division, Fort Campbell, KY presented me with the DFC. Colonel Bush did not forget.

The Fort Campbell Courier

Chief Warrant Officer Michael Gulick with wife Lynn Gulick receiving the DFC

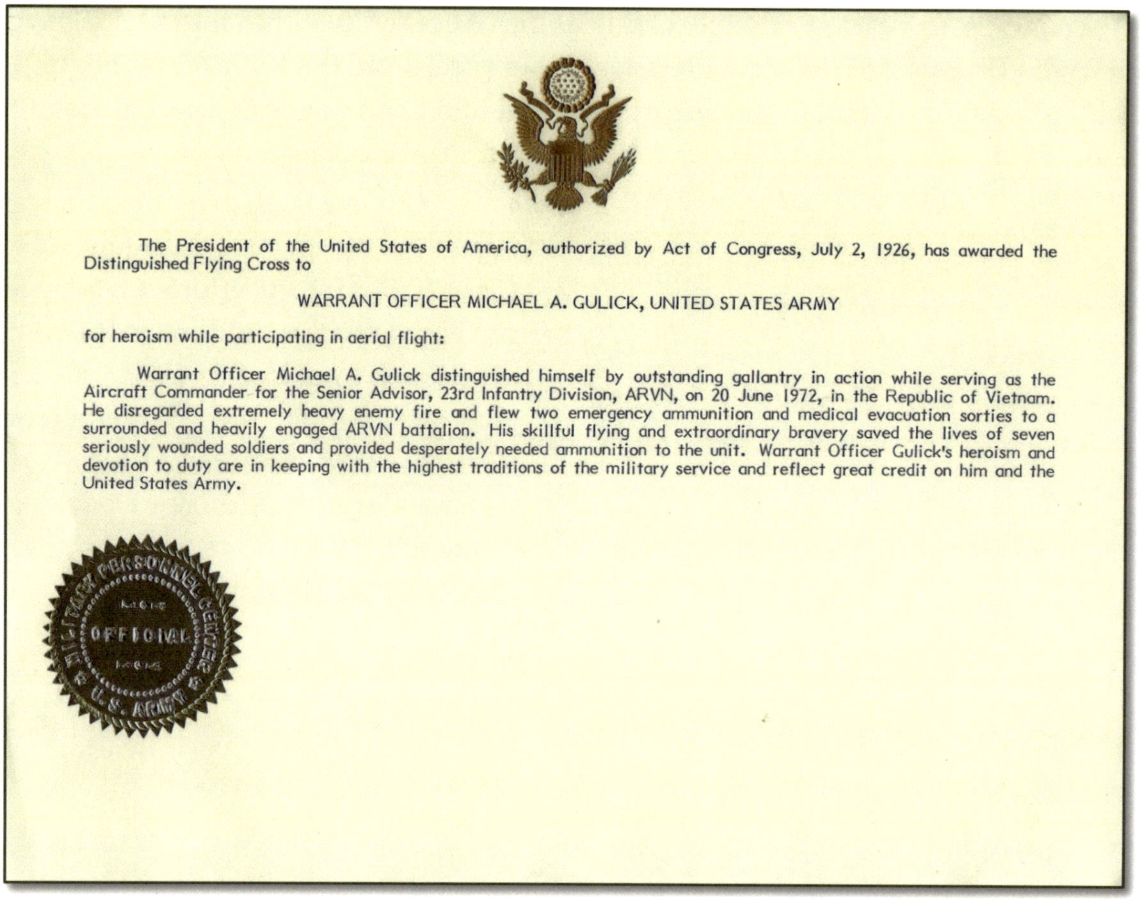

The President of the United States of America, authorized by Act of Congress, July 2, 1926, has awarded the Distinguished Flying Cross to

WARRANT OFFICER MICHAEL A. GULICK, UNITED STATES ARMY

for heroism while participating in aerial flight:

Warrant Officer Michael A. Gulick distinguished himself by outstanding gallantry in action while serving as the Aircraft Commander for the Senior Advisor, 23rd Infantry Division, ARVN, on 20 June 1972, in the Republic of Vietnam. He disregarded extremely heavy enemy fire and flew two emergency ammunition and medical evacuation sorties to a surrounded and heavily engaged ARVN battalion. His skillful flying and extraordinary bravery saved the lives of seven seriously wounded soldiers and provided desperately needed ammunition to the unit. Warrant Officer Gulick's heroism and devotion to duty are in keeping with the highest traditions of the military service and reflect great credit on him and the United States Army.

DFC award to Michael Gulick for actions taken in 1972 in Vietnam

I returned to the States in August 1972 with an assignment to the 205th Aviation Company (CH-47 model C), Fort Benning, Georgia as a helicopter pilot. Once back, I started attending night school to earn a college degree. In September 1972, our unit was selected for deployment to Germany. We arrived at Finthen Army Airfield in Mainz, Germany and I remained there for the next five years, flying the CH-47C helicopter. I was promoted to Chief Warrant Officer (CWO-2) in 1972. I completed my Bachelors' of Science in Aeronautical Science from Embry-Riddle Aeronautical University, Daytona Beach, Florida in 1976.

In January 1977, I was assigned to B Company, 159th Aviation Battalion, 101st Airborne Division, Fort Campbell, Kentucky. While there was promoted to Chief Warrant Officer 3 (CWO-3). A little while later I applied for and was granted a Direct Commission to 1st Lieutenant (O-2) in the Transportation Corps on 25 January 1980.

Promotion to 1st Lieutenant

My first assignment as a Commissioned Officer was to Howard Air Force Base, Panama Canal Zone, Panama as the Maintenance Platoon Leader, 590th Aircraft Maintenance Company. I then moved to the Battalion Staff as the S4 (Supply & Logistics) Officer where I was promoted to Captain (O-3).

I completed fixed wing training en-route to my new assignment at the Army Aviation Engineering Flight Test Activity, Edwards Air Force Base, California on May 1984.

We used two of the Navy's Fixed wing training aircraft as "chase" planes to follow the Army aircraft that were being flight tested. Normal Army fixed wing aircraft could not effectively perform the "chase" mission as the "Chase" plane needed to be faster than the test aircraft. I was one of a few Army Aviators to be qualified in these aircraft.

Beechcraft T-34 Mentor at Edwards Air Force Base, CA

North American T-28 Trojan at Edwards Air Force Base, CA

Canadian UH-60 undergoing Icing test

CH-47D Helicopter Ice spray system

CH-47 Helicopter Ice Spray System (HISS) testing the V22 Osprey

In order to fully test the ice capabilities of aircraft, we needed to do so in the safest manor possible. To accomplish this the Flight Test Activity modified a CH-47D helicopter to carry a water spray rig that could be deployed in flight. A 600-gallon water tank was fastened in the cargo area, water pumps and electronics were added. Testing would be on in Duluth, Minnesota during the winter months. The test aircraft would fly behind the CH-47 and when ready, would fly into the water spray. The water would immediately freeze on the flight surfaces. If the aircraft starting experiencing flight issues, then were able to fly out of the spray and then safely land the aircraft. All services used this system for Icing Verification.

Soon after, I started attending night school for post graduate studies and in 1985 I completed a Master's of Business Administration from Golden Gate University, San Francisco, California.

February 1988 saw me heading back to Germany once again, but this time as a Deputy Material Fielding Officer for the deployment of the CH-47 Model D helicopter to all units in Europe. There I was promoted to Major (O-4). Soon after the Desert Storm conflict with Iraq began.

Instead of taking command of the Materiel Fielding Teams in Iraq for the deployment of Army aircraft to that region, the Program Executive Officer (PEO) for Aviation requested me to report to St. Louis, Missouri as the Executive Officer/R&D Coordinator, affectionately known as the "General's Aid", to the PEO – Aviation, Mr. Gary L. Smith.

I was subsequently reassigned to St. Louis in July 1990. Here, I supported the Army's acquisition process in support of Operation Desert Storm. An overview of some of our work was published in the Army Aviation Magazine.

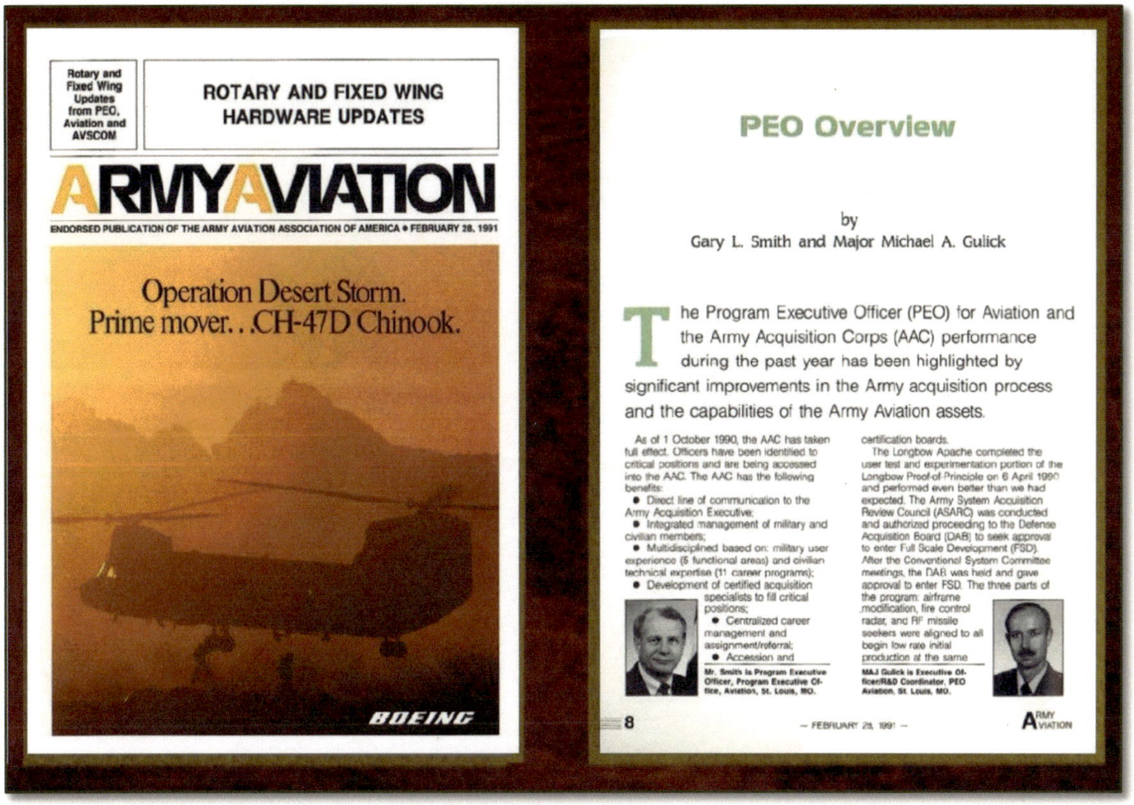

PEO Overview

by
Gary L. Smith and Major Michael A. Gulick

The Program Executive Officer (PEO) for Aviation and the Army Acquisition Corps (AAC) performance during the past year has been highlighted by significant improvements in the Army acquisition process and the capabilities of the Army Aviation assets.

As of 1 October 1990, the AAC has taken full effect. Officers have been identified to critical positions and are being accessed into the AAC. The AAC has the following benefits:

- Direct line of communication to the Army Acquisition Executive;
- Integrated management of military and civilian members;
- Multidisciplined based on: military user experience (5 functional areas) and civilian technical expertise (11 career programs);
- Development of certified acquisition specialists to fill critical positions;
- Centralized career management and assignment/referral;
- Accession and certification boards.

The Longbow Apache completed the user test and experimentation portion of the Longbow Proof-of-Principle on 6 April 1990 and performed even better than we had expected. The Army System Acquisition Review Council (ASARC) was conducted and authorized proceeding to the Defense Acquisition Board (DAB) to seek approval to enter Full Scale Development (FSD). After the Conventional System Committee meetings, the DAB was held and gave approval to enter FSD. The three parts of the program: airframe modification, fire control radar, and RF missile seekers were aligned to all begin low rate initial production at the same

Mr. Smith is Program Executive Officer, Program Executive Office, Aviation, St. Louis, MO.

MAJ Gulick is Executive Officer/R&D Coordinator, PEO Aviation, St. Louis, MO.

8 — FEBRUARY 28, 1991 — ARMY AVIATION

Published PEO Overview

I wanted to retire in the uniform I was issued in Basic Training, as a testament to all the experiences, lessons learned, and growth the Army and the uniform had afforded me since that day I reported to training back in 1970, and did so when I retired from the Army in April 1992, after 22+ years of Active Duty.

Major Michael Gulick at time of Retirement from the Army in 1992,
In the uniform issued to him in Basic Training in 1970

After retirement, I stayed on supporting the Army as a civilian contractor for the next 6 years. In the end, I completed 28 years of service to the Army.

After the Army, I worked in various civilian roles, including Assistant Vice President and Global Lead of Problem Management for an International bank. Retiring from that bank after 15 years of service. I am now fully retired and am enjoying life in The Villages, Florida.

Michael Gulick today in retirement

19

James A. Howd
Captain
United States Army (Ret)

Vietnam War

A young Lieutenant Howd

The date was 17 April, 1968—a beautiful spring day in the tropical paradise of Cu Chi, South Vietnam. If my memory serves me correctly, I was (at the time) the Operations Officer for "A" Company (the Little Bears), 25th Aviation Battalion, 25th Infantry

Division. Our company commander was MAJ Roger McLeod, and our Company Executive Officer was CPT Jim Jaap. On that date I had flown as the Aircraft Commander and flight lead on a mission involving some 15 or 20 Huey helicopters and crews. We had just returned to our POL (Petroleum, Oil and Lubricants)[3] to top off (refuel) after completing a successful and (fortunately for us) uneventful multiple ship combat assault. The air crews were all hot and tired, but glad to be back in Bear Country safe and sound following the Infantry Unit insertion that we had executed without incident. For all of the Little Bear crew members it was just another day of safe flying, culminating in a successful mission and a happy return to our base camp while (thankfully) suffering no casualties. And then came the radio call…

Because I was in the lead aircraft for the combat assault, my aircraft was the first to land at the POL point to refuel. We had topped off and I was preparing to pull pitch to return the ship to the Bear Pit—our unit's helicopter parking area made up of a number of PSP (Perforated Steel Planking) and dirt revetments that provided protection to our ships from incoming mortar and rocket fire—when I received a radio call from CPT Jaap on the company FM radio. Seems that a LRRP (Long Range Reconnaissance Patrol) Platoon consisting of some 13 men was in deep sh_t to our north. While on recon patrol they had come in contact with a greatly superior force of Viet Cong (VC) and North Vietnamese Army (NVA) soldiers, and the LRRP's were pretty much pinned down by some intense small arms fire. Word was that the LRRP's needed to be extracted ASAP or face the strong possibility of annihilation. CPT Jaap told me that he and MAJ McLeod would be flying a Command and Control (C&C) ship—that I needed to dispatch two aircraft and crews to the north for the purpose of conducting an emergency extraction—and that the C&C would marry up with the extraction ships en route to the proposed PZ (pickup zone).

I looked towards the POL point to my immediate right and saw that CPT Stephens and his crew had completed their refueling and were just beginning to pull pitch for their return to The Bear Pit. I keyed the mike, called CPT Stephens, and the two of us pulled out and headed north to rendezvous with the C&C at or near the LRRP Platoon's position. Mine was the lead aircraft, with CPT Stephens and his crew flying as my right wing.

Within a matter of minutes, we had arrived in the target area where the LRRP's were pinned down – C&C had instructed the LRRP's to pop smoke (to identify for us their location on the ground) – and our two ships began lining up for the approach into the PZ. CPT Stephens, to my right rear, was to take the LRRP's on the right and I would recover the elements on the left. CPT Stephens wasted no time, and as soon as he spotted the smoke on the right, he made a rapid approach in the direction of his smoke. He passed me on my right

[3] The POL point was the Vietnam combat equivalent of a gas station: 10 huge fuel bladders in open dirt berms with a few nozzles we used to fill our fuel tanks.

and within a few seconds I heard over my radio his blades chopping their way through the trees. At the same time CPT Stephens called out "trees, trees!"

Too late for him as he had literally chopped his way into the PZ, barely touched his skids to the ground, his six troops leaped onto his Huey, and he was pulling pitch and getting out of there. He was in and out in what seemed like a matter of seconds. We, on the other hand, identified our smoke and began to descend into the pickup zone. As we approached the spot it became obvious that our LRRP's were in an area where there had been a fire, and the fire had burned all the foliage off the trees but left the trunks sticking straight up into the air. We were approaching an area with small sapling trunks sticking straight up at different heights, a situation which I concluded made it impossible to land the chopper. So, we came to a high hover, set the belly of the bird on the top of the trees, and the troops on the ground began their climb—hand over hand—up the burned trees to the underbelly of the helicopter where they grabbed the skids and were pulled aboard by the crew chief and the door gunner.

While hovering above the trees I was afraid my tail rotor would strike a tree and we would crash, so I instructed the crew chief and door gunner to "Keep me advised of our tail rotor location in relation to the trees, and keep the tail rotor out of the trees." Meanwhile, the guys on the ground were climbing – hand over hand – up the trees, making their way to the underbelly of the aircraft where they could grab the skids and be pulled into the cargo compartment of the aircraft.

While hovering above the trees I was afraid my tail rotor would strike a tree and we would crash...

The C&C advised us that we were receiving some pretty heavy small arms fire, and that we should expedite our departure. I was totally consumed with my efforts to maintain a steady hover and keep both the tail rotor and the main rotor blades clear of the trees. I did a quick look to the front and saw no tracers, so assumed we were OK and continued the hover. In the meantime, the crew chief and the door gunner continued to update and reassure me that the tail rotor was OK, and that the troops were making headway in their climb up the trees.

Grabbing the skids

Again, C&C called and told me to expedite our departure – that heavy small arms fire was all around the aircraft – and I responded that we could go no faster than the LRRPS could climb. Then, before all were aboard, the call came in to abort – to pull out before I got the aircraft shot down. I responded: "As soon as the last man is aboard, we will be out of here." At about that time either the crew chief or the door gunner (and I do not recall which one made the call) frantically announced – "all men aboard – tail rotor clear – let's go!" And we pulled up and away for our return to Cu Chi.

Upon arriving at Cu Chi we landed near the CG's hot spot to off load the troops. An ambulance pulled close as we hovered to a stop. One of the LRRP's (who had been shot in the leg / knee while climbing up the tree to grab the skids) literally fell out of the aircraft onto the ground, pulled off his helmet, and kneeled to kiss the earth. I was standing there just outside the ship with my helmet still plugged into the communication system when he arose, looked me in the eye, and thanked me for saving their lives.

Two things I considered significant came out of all this.

First – a couple of months after the extraction I was awarded the DFC for heroism in connection with this extraction.

Citation:

> "CPT James A Howd Jr. is awarded the Distinguished Flying Cross for action on 17 April 1968 for heroism while participating in aerial flight: CPT Howd distinguished himself by heroic actions on 17 April 1968 while serving as aircraft commander of a UH-1D helicopter for Company A, 25th Aviation Battalion on a combat operation in the Republic of Vietnam. A long range reconnaissance patrol was in heavy contact with a numerically superior enemy force and Captain Howd's aircraft was called upon to help extract the team. With complete disregard for his own safety, CPT Howd directed his helicopter into the intense enemy fire. One patrol member then managed to leap up, grab a skid, and pull himself into the aircraft. One by one, the remaining soldiers climbed aboard. Once they were safely aboard, CPT Howd skillfully maneuvered his aircraft out of the extremely hazardous landing zone, taking such effective evasive action that his aircraft was never hit by the intense enemy fire. His valorous actions and outstanding leadership contributed immeasurably to the success of the mission. CPT Howd's personal bravery, aggressiveness, and devotion to duty are in keeping with the highest traditions of the military service and reflect great credit upon himself, his unit, the 25th Infantry Division, and the United States Army."

They (the Army, my superiors, etc.) thought I was brave to continue to hold the aircraft at a hover, waiting for the last of the LRRP's to climb up to and board the aircraft, while receiving intense enemy fire. The reality was that I wasn't being brave at all. I was actually blind as to what was happening with the enemy fire. I had looked, saw no tracers, and assumed that we were OK. Fact is I was blinded by the bulkhead of the aircraft – not brave.

Second – I retired from the Army in 1976. Some nine years after the LRRP extraction I had the opportunity to walk past a man at Ft Campbell, KY. He was limping as we approached each other and as we passed, I had the distinct feeling that I knew him from somewhere in my past. Each of us had taken a few steps, stopped, then turned and looked at each other. Suddenly he began to approach me and I him. As it turned out this man was the soldier who was wounded in the leg and who got out of the aircraft and kissed the ground after we landed at Cu Chi back on 17 April 1968. We both looked considerably different in 1977 (than we looked in 1968), but we both recognized each other almost immediately – after all that time – and with but seconds of actually seeing each other back in 1968. It's a small world we live in, but one filled with wonders.

Captain Jim Howd, US Army (Ret)

20

Phillip Ide
Major
United States Army (Ret)

Vietnam War

Phillip Ide at 20 years old in Vietnam

The stories behind 4 Distinguished Flying Crosses while flying for the 1st Air Cavalry Division (Airmobile) 1967 - 1968

27 August 1967

I was assigned the mission of logistic supply for the 3rd Brigade, 1st Air Cavalry. The logistical mission required that I would be responsible for aerial support of this unit, which included ammunition, food, water, troop replacements, or possible medivacs, regardless of the weather, enemy situation, or landing conditions at the Landing Zone (LZ).

Bell UH-1 at an LZ at dawn

This particular rainy morning, the clouds were hugging the mountain tops, and to add to the mission complexity, the LZ was a small clearing which had been opened by a bomb onto the top of relativity high ridgeline. The infantry Battalion that I was supporting was maneuvering in the Soul Ca Valley, adjacent to the Ia Drang Valley, of **WE WERE SOLDIERS ONCE…AND YOUNG** fame.

Upon arrival at the ridgeline LZ, the first landing was uneventful. The back edge of the LZ dropped off several hundred feet. We had to stick the nose of the aircraft into a small opening surrounded by 300′ trees, allowing just enough room for the nose and the rotor blades to fit. We landed, with our tail hanging over the edge of the valley below, unloaded, hovered out backwards, pointed the nose down the ridge-line and flew away, just clearing the treetops.

After first light, an infantry company started maneuvering down the ridgeline with the objective of driving any enemy forces headlong into armored elements of Delta Troop, 1/9th Cav. located on a road at the base of the mountain. Ideally, this pincer would manifest itself on or near the road that meandered through the valley, were heavy weapons of the armor unit would dominate the battle.

As the infantry company maneuvered down the ridge-line, lead elements ran headlong into a bunker complex that was well fortified and manned by an enemy force armed with heavy machine guns that engaged the point elements. Quickly realizing that he was outgunned, the Company Commander backed off from the frontal attack and started to regroup and move back up the ridge-line. During this maneuver, his entire company was surrounded and a major firefight broke out!

The only logical escape route was to fight their way back uphill to the LZ if they were going to survive! Attacking downhill into the well-fortified bunker complex would be suicidal!

As the battle raged on, the Battalion Commander, ordered me & my wingman to start dropping ammunition through 300' high foot trees, to the infantry company below! The only way to locate them beneath the sea of trees was to request a smoke grenade be launched and wait for the smoke to drift upwards through the trees. Once the smoke color was confirmed, I would position the helicopter over the drifting smoke, as the crew chief and gunner threw boxes of ammunition overboard. Hovering on the side of the side of a mountain at an extremely high hover uses every bit of engine power that the Huey has and it takes good piloting skills to stay in place over the drop area and not spin out of control!

During the first ammunition drop, we had our flight control hydraulics shot out and had to vacate the mission. We were given a different helicopter and good luck wishes. On the second ammunition drop, my wingman had his copilot shot in the leg, which unnerved us! On the third run, we had an infantry Sgt. onboard shot, while assisting with the throwing of ammunition boxes overboard.

Despite the harassing enemy small arms fire, we continued dropping ammunition and water throughout the afternoon into dusk. To further complicate our resupply mission, we had to operate perilously close to a constant barrage of supporting artillery fire. In order to maneuver on the mountain side to begin the unloading process, or to vacate the drop zone, we had to continually overfly the enemy and they never missed that opportunity to shoot at us! Once darkness enveloped the battle area, we stayed on station, carrying ammunition and evacuating the injured from the difficult landing zone that was improving thanks to some combat engineers.

Each night approach to the LZ was made to the light of handheld flashlight wands held by a pathfinder. We were unable to use our own high-powered search light because of enemy fire, making the final approach to the LZ quite interesting (terrifying)! Around midnight the infantry company finally closed around the LZ and the Bn. Commander released us, with many thanks!

At first light our soldiers were successfully extracted from the ridgeline by, "C" 227th Assault Helicopter Company.

16 October 1967

"C" Co, 227th Aviation Battalion, was cited for the periods 01 to 31 Oct 1967, for the most **Valorous Unit Citation (VUC)**, which read:

> "Flying in the mountainous terrain of the Que Son Valley, with its inherent clouds and heavy rainfall, made low level, contour flying through the valley a necessity. Ceiling and visibility were often reduced to zero, zero. Anti-Aircraft fire was encountered on almost every combat flight operation causing "C" Co, 227th Aviation Battalion to lose several aircraft to enemy fire."

On 16 October 1967, I was part of a Ready Reaction Force (RRF) and was assigned the position of White 1, the next to last aircraft in a formation of six UH-1H helicopters

Our mission was to support the 2nd Bn, 12th Cavalry, 1st Cavalry Division (Airmobile) - which was operating approximately 25 nautical miles northwest of Chu Lai, I Corps. We had been conducting a battalion size troop move when our flight leader, CPT James Lawton received a frantic radio call that the 2/12 Command and Control ship had been downed by enemy fire with the battalion leadership on board and their condition was not known. Both pilots were wounded.

It was decided to carry 35 ready reaction (RRF) troops to the crash site. The senior officers felt that they would be able to lay down a volume of supporting fire and secure the immediate area around the downed helicopter. We were going to need extra time on the ground to load the wounded from the helicopter, located in a rice paddy with waste deep water.

Before departing, CPT Lawton radioed the tactical situation to our supporting gunship team leader. Recognizing the gravity of the situation and the necessity for haste, CPT Hudgins, of D, 227th, elected to disregard basic gunship tactics and accompany our flight as a single gunship. He did this with the knowledge that he would have no cover for his Attack Helicopter while doing gun runs against enemy threats. En route to the LZ, we learned that the aircraft on the ground had received hits from both .51 and .30 cal automatic weapons and that the LZ was HOT!

The crash site was a large rice paddy, with the downed helicopter located in the center. The ceiling was approximately 500' absolute with visibility periodically impaired by mist and rain showers. Having no direct contact with the downed aircraft and unable to determine the direction of enemy fire, our flight leader elected to land into the wind and surround the downed aircraft with our helicopters to cover the evacuation process.

Flight of UH-1 Hueys approaching an LZ
Geysers on the right suggest suppressing fire to cover their approach.

As soon as our flight of six touched down, we began receiving automatic weapons fire. Rounds were splashing in the water between and around our helicopters. We were seeing multiple muzzle flashes from automatic weapons in the tree-lines on both sides of our sitting flight. Our gunners, the inserted infantry and the one gunship kept returning fire degrading their accuracy. Squad size North Vietnamese Army (NVA) elements tried advancing toward us, as we sat exposed loading the wounded. I tucked my legs up and tried to get very small behind the seat armor plating! My gunner was literally shooting at enemy soldiers 200 meters away as they tried to close onto our position. His shooting skills were verified by the sight of dead enemy soldiers lying near the downed helicopter, as we lifted off!

The covering gunship was forced to repeatedly overfly automatic weapons positions in order to lay suppressive fire on all the enemy fire reported. Because of the delay in loading the wounded personnel, the gunship expended all their mini gun ammunition and rockets, with the exception of a hand held M-79 grenade launcher.

On departure, we came under fire from a twin 51 caliber anti-aircraft weapon. The gunship attacked, in a last-ditch effort, at a very low level. The anti-aircraft position and its crew was taken out by a well-placed M-79 round fired out of the co-pilot's window destroying the threat. (Silver Star Material)

Although our flight was subjected to constant enemy fire on both lifts, the Battalion Commander, flight crew and the RRF force were all successfully evacuated.

Unfortunately, less than month later, the same 2/12th Bn. Commander, LTC Gregory, ordered his Command & Control ship to once again overfly an enemy position at low level, and all were lost!

31 January 1968

On January 31, 1968, amid the celebration of the Vietnamese New Year, the enemy launched the TET Offensive, a major effort to overrun South Vietnam. Some 10,000 enemy, well equipped, crack NVA regulars blasted their way into the imperial city of Hue, overpowering all but a few pockets of resistance held by ARVN troops and the US Marines.

The morning of January 31st, I was one of six flight crews that was ordered to pack our personal gear and fly from our base in Chu Lai to a Sea Bee compound just outside of the city of Hue. We were to spend the night there, then proceed the following morning to the new 1st Air Cavalry Firebase at Camp Evans, 20 miles north of Hue.

The only intelligence briefing we received before leaving, was that there was a lot of enemy movement around Hue. We needed to be in position to support the lead elements of the 3rd Brigade that was in the process of being moved north to Camp Evans by truck and C-130 from An Khe, on 1 February, 1968.

The beginning of the early morning flight was beautiful and uneventful until we approached the Perfume River that flows through the heart of the city of Hue. Approaching the river, we were greeted by bursts of flak, and green tracers that lighted up the sky! We immediately dove for the tree tops and safety from the big AA guns!

During our escape from the anti-aircraft guns over Hue, we flew low level over a large cemetery that literally had a NVA soldier behind every grave stone! We learned later that they were gathering pre-stored weapons hidden in grave-sites for the night attack on the city. They were in the process of arming a large number of soldiers that had walked into Hue with weapons undetected.

As we roared over their heads, I took a round through my door that cut the sleeve on my shirt. That got my attention! Welcome to Hue and TET in '68!!

Welcome to Hue and TET '68!

We refueled and licked our wounds at Marine Hue, Phu Bai, and repositioned the helicopters for the night on a softball diamond located on the Sea Bee base. Later that afternoon, as we were enjoying ice cream in the mess-hall, mortars started to explode nearby, driving the See Bees onto the floor. We had to step over them to get more ice cream and we all realized at that "moment," that we had been in combat too long and being mortared had become just another routine event! The Sea Bees stared at us, as we continued to eat our ice cream, like we were some kind of crazies!

Around 9:00 PM, CWO Fred Ferguson and I decided to share a bunk bed. I took the top bunk and he the bottom, and we settled in for the night. (Fred would be nominated for the Medal of Honor for his actions the next morning in the city of Hue.)

Around 02:00 a rocket blew up the building next to us and the concussion from the blast was so strong that it blew the bunk bed over with us in it. In the darkness and confusion, we managed to grab our weapons, boots and head for the bunkers, trying not to hurt ourselves in the melee. The mortar and rocket attack was intense and word spread that the enemy was close to breaching the perimeter! Observing the confusion around us, especially in the ranks of the non-combat, unarmed naval personnel, I decided to run for my aircraft and attack the bad guys from the air. I had already been on two firebases where Sappers had broken through and attacked us, and I did not want another repeat performance on unfamiliar fighting ground!

Tracer rounds visible during night flight

I grabbed my crew chief and together we ran through the falling mortars and made it safely to the helicopter. It was surreal seeing the mortars explode near us, just like in the movies! I strapped myself in and I took off single pilot with only my crew chief onboard!

Once airborne, I maneuvered the Huey to a position over the heads of the attacking enemy and my crew-chief with his M-60 machine gun, fired directly into the enemy positions from 200' over their heads. The surprise attack from above, broke the attack momentum of the attacking enemy force and I like to think kept the base from being overrun! We took several hits in the aircraft fuel bladders and other than losing fuel, the helicopter flew fine.

CWO Fred Ferguson and crew departed 30 minutes later and we joined up over Hue and landed at Marine Phu Bai for fuel. We spent the rest of the night in the passenger terminal that had been attacked prior to us arriving. The big plate glass windows in the passenger terminal were all shot out, and broken glass was everywhere. We formed a perimeter in a good fighting position and waited for the sun to rise!

The 1st full day of TET was just the beginning of 21 days of helicopter mid intensity war operations while flying in Monsoon weather with enemy soldiers seeming everywhere shooting at us!

The day was going to be quite momentous for Ferguson. It was during this first day of the battle for Hue, that he would distinguish himself and be awarded the Congressional Medal of Honor!

The 1st Air Cavalry Invades the A'shau Valley

19 April 1968

Commanding General John Tolson writes:

"North Vietnamese long range 23mm and 37mm antiaircraft guns, a rarity in South Vietnam, and 50 caliber machine guns destroyed or damaged about 30 1st Cavalry helicopters in the first 24 hours of the raid. Some of them were hit at 6,000 feet! Hell, I've never lost that many helicopters in weeks and weeks. By far it's the hottest place we've ever gone into and the most losses we've taken in a single day," exclaims MG Tolson!

On 20 April, the first day of the assault into the A'Shau, Major Burkhalter, our Company Commander, led "C" company's yellow and white flights for an assault on a ridge line four (4) kilometers southwest of LZ Tiger. LZ Tiger had been established earlier in the day by the 229th, Assault Helicopter BN, taking huge helicopter losses. The LZ for us was a bomb crater surrounded by 300-foot trees, barely acceptable as a LZ, but we were ordered into it by the C & C bird. The weather at this time was very marginal and we could see that a second insertion would probably be impossible.

Yellow 1's approach was made along a ridgeline then he dropped down into the bomb crater studded with stumps and debris. Yellow 1 elected to come to a high hover, experienced power loss and crashed. The crew was not injured, but had to remain on the ground for several days evading enemy forces until they finally linked up with the Air Cavalry near LZ Tiger. Rescue was impossible, the weather moved in directly after the crash and all rescue attempts had to be abandoned and all troop lifts suspended, leaving the few American troops inserted into the A'Shau abandoned and grossly outnumbered.

The second day of the invasion I was Yellow 3. Yellow 1 dropped off his troops with difficulty and departed, while the second helicopter in line attempted to come to a high hover, lost power and crashed, sliding down a slope and ended up at the bottom of the bomb crater.

I was next and successfully found a flat spot high up on the crater and placed a landing skid onto it. We successfully dropped off our troops and waited long enough to extract the uninjured crew from the crashed helicopter. While climbing vertically out of the bomb crater an enemy soldier stepped out from behind a tree, literally looked me in the eye, took aim from 50' away, and the AK-47 failed to fire, a second later we were gone! Our six helicopters lined up in trail behind us and managed to insert their load of troops onto the LZ called Pepper before the weather halted all flight operations.

At dusk I volunteered to try to evacuate what was left alive of a Chinook crew that had been shot down in the Valley trying to deliver an artillery piece. The crash site was on the side of a 4000' mountain overlooking the A'Shau valley. The Chinook, luckily, had crashed into an open area large enough to make the rescue possible without a hoist. I was given a radar vector penetration from the GCA site at Camp Evans into the valley, then had to look for signs of the downed Chinook, which turned out to be a flashlight held by a crewmember.

Once we saw the light, I had to hover down through an opening in the jungle that was surrounded by trees. I did this in the dark without landing lights only guided by an injured crew-member holding a flashlight to mark the landing area, if you wanted to call it one!

Fortunately, the enemy anti-aircraft gunners could no longer see me, only hear us coming and going, so their fire was inaccurate. Just scary! Knowing that their fire was inaccurate gave us the opportunity to be able to come to a high hover over the trees, stabilize, and then be able to descend vertically through the hole in the jungle and disappear from sight. The soft red glow from parts of the burning wreckage made the rescue scene surreal and it did give me some minimal light to land by.

After loading the four injured crewmembers, I initiated a climb to 6,000 ft., to clear the mountains. I contacted the GCA site and made an IFR descent through the rain into Camp Evans. It was really a hairy rescue, but one to be proud of!

During that day, "C" 227th lost two aircraft that were not recovered and most of the helicopters that flew into the A'Shau those two days in April had battle damage.

22 April brought much improved weather conditions and the airlift into a much improved LZ Pepper continued with all of our flyables committed. Anti-aircraft fire was harassing but far less than previous days and gunships were everywhere. At the day's end the 3rd Brigade was firmly entrenched into the A'Shau Valley.

Gerry Owens!!

Phillip Ide has over 40 years of aviation experience in a career heavily involved in regulatory control and oversight.

After his Vietnam tour with the 1st Air Cavalry in Vietnam, Phil was assigned as an Instructor Pilot for the Army at Fort Wolters, Texas, teaching in the TH-55 basic helicopter trainer. Upon release from Army active duty in 1970, he started his civilian aviation career flying a Piper Aztec modified for aerial photography and mapping. It belonged to a mapping and surveying firm in Watertown, NY. He then moved Burlington, VT and flew for Air North, a Regional Air Carrier.

In 1972 he was hired as a Department of the Army Technician Flight Instructor for the New York Army National Guard, in Albany, NY. He instructed in all of the UH-1 helicopters. Being a dual-rated Army aviator, his instructor pilot's duties also included the U-8F (Queen Air) and C-12D (King Air)

In 1980 Phil was selected to be the Commander of the Army Aviation Flight Support Facility, Albany, NY. During his tenure as Commander, he was appointed Military Aviation Officer for the 1980 Winter Olympics, in Lake Placid, NY. There he worked closely with the Olympic Committee, the Secret Service and the FBI. He also oversaw several aviation emergency operations within New York State, to include floods, snow emergencies and forest fires.

Other important National Guard assignments included the introduction of the AH-1S Attack Helicopter and the C-12D King Air into the Army National Guard System. He helped form a Military Aid to Safety and Traffic (MAST) for rural hospitals transferring patients in the Adirondack and Catskill Mountains areas to Albany Medical Center, Albany, NY.

One of his more exciting assignments was to lead a flight of six C-12F reconnaissance airplanes from Moffett NAS, California to Tel Aviv, Israel, as part of the Camp David Accords.

In 1987 Phil decided on a career change and left the Army Aviation Flight Support Facility and joined the FAA Flight Standards Service. He still remained a National Guard Reservist until retirement at age 60. He logged over 38 years of military service and over 8500 flight hours.

As a FAA Aviation Safety Inspector, he was responsible for the oversight of Federal Aviation Regulation compliance. Phil was the Principal Operations Inspector for Chautauqua Airlines, a major Regional Airline that moved from New York to Indianapolis and became Republic Airlines, a United Airlines partner. He also had regulatory responsibility for several FAR 135 Air Taxi operators and a foreign based Flight Safety International Training Center in Toronto, Canada.

Phil served several years as the FAA Office Manager and Assistant Manager for the Rochester Flight Standards District Office. The FSDO oversees 40 FAR 135 operators, 45 Repair Stations, 80 FAR 129 Foreign Certificates and the nation's second largest fractional share company, Flight Options.

As a FAA Manager he has taken a leadership role in audits of FAR 135 Emergency Medical Service Operators. He worked closely with a special team from the FAA National Headquarters, Washington DC, to help improve EMS Operational Control Oversight and improve the dismal accident rate in that industry.

After his FAA retirement he was employed as an aviation consultant with JDA, an airline Certification and Compliance business, Washington, DC. In 2011, Phil fully retired from aviation and now divides his time between Rochester, NY and The Villages, FL.

Major Phillip Ide at an event for DFC recipients

21

Wayne Paul Lotsberg
Lieutenant
United States Navy

Vietnam War

Lieutenant Wayne Paul Lotsberg, USN, 1963

Wayne Lotsberg had 374 combat missions over North Vietnam during four cruises in the Navy. He had a cruise on USS Ranger, USS Constellation and two aboard USS Midway. He has been awarded three Distinguished Flying Crosses (DFCs) and numerous other awards for combat heroism. In between sea tours, he served as a Flight Instructor in the Navy Advanced Flight Training Command at Kingsville, Texas. He has flown the A4 Skyhawk, F4 Phantom, and A7 Corsair II combat aircraft as well as several training aircraft. Wayne was also a certified Landing Signal Officer (LSO) whose duty is to "coach" via radio, the pilots landing on the carrier, especially at night and during adverse sea or weather conditions.

Wayne enlisted in the Navy in November, 1962, fresh out of high school. During boot camp in San Diego, he was given several aptitude tests and interviews and was accepted to Flight Training via the Naval Aviation Cadet (NAVCAD) program. At age 19, he began flight training in Pensacola, Florida, achieving his first carrier landings in the T28 Trojan at age 20, then flying the F9 Cougar and F11 Tiger earning his Navy Gold Wings in April 1965. Photo above taken in 1963 at age 19.

After training in the A4 Skyhawk, he deployed aboard USS Ranger in December 1965 for his first cruise to Vietnam. Most missions in his first two cruises were bombing runs in what has been called "The most heavily defended real estate in the world". Many dive-bombing runs were amidst AAA (Anti-Aircraft Artillery) so thick, the airplane was jostled about by the percussion of the AAA bursts. In his first two cruises, he had 240 combat missions and had been awarded his first two DFCs for combat heroism and bravery.

Douglas A-4 Skyhawk

The first was for bombing an important, heavily defended railroad bridge.

Citation:

> "The President of the United States takes great pleasure in presenting the DISTINGUISHED FLYING CROSS to Lieutenant (junior grade) Wayne Paul Lotsberg for heroism and extraordinary achievement in aerial flight as the pilot of an A4E Skyhawk while attached to Attack Squadron FIFTY-FIVE embarked in USS RANGER (CVA 61) on 8 May 1966. Lieutenant (junior grade) LOTSBERG distinguished himself by his courage and skill while attacking an important railroad bridge in the heavily defended Tho Trang marshaling yards. He pressed home his accurate attack despite the intense enemy AAA and automatic weapons fire directed at his aircraft. Dropping his six 500-pound bombs with lethal accuracy, he demolished the bridge. He superbly demonstrated courage and skill and deprived the enemy of a vital link in their rail transportation system. Lieutenant (junior grade) Lotsberg's gallant and professional actions were in keeping with the highest traditions of the United States Naval Service."

The second DFC happened as the later portion of a bombing strike when he was called upon to assist in the rescue of a downed pilot.

Citation:

> "The President of the United States takes great pleasure in presenting a gold star in lieu of a second DISTINGUISHED FLYING CROSS to Lieutenant (junior grade) Wayne Paul Lotsberg for heroism and extraordinary achievement in aerial flight as the pilot of a jet aircraft attached to Attack Squadron FIFTY-FIVE embarked in USS CONSTELLATION (CVA 64), on 3 October 1967. Lieutenant (junior grade) LOTSBERG was the section leader in a division of A4C SKYHAWKS during a major strike against the strategically important LOC BINH highway bridge southeast of Lang Son, North Vietnam. In an outstanding display of singular weapons delivery skill he placed all his bombs precisely on target completely destroying one span of the bridge. Retiring from the target area, Lieutenant (junior grade) Lotsberg was informed that a pilot had been shot down north of Haiphong. After inflight refueling his aircraft, he proceeded to the area of the downed pilot. In the face of continuous enemy fire he conducted a search at low altitude until locating and establishing contact with the pilot. Lieutenant (junior grade) Lotsberg's eagerness to help others despite great personal risk and his superb airmanship under fire were in keeping with the highest traditions of the United States Naval Service."

Taking his combat experience and LSO qualifications to Texas, he was able to share his experience and utilize his teaching abilities to train students to meet the very high standards of being a Naval Aviator. He taught all phases including Air Combat Maneuvering (ACM) and Air to Air Gunnery, as well as qualifying students in carrier landings.

Returning to combat in Vietnam in 1971, he was one of the most experienced and respected combat pilots and served as a mentor to junior aviators flying the A7 Corsair II and flew another 134 missions. Because of his experience, he more often was designated to lead "SAM" (Surface to Air Missile) suppression flights. This entailed enticing the SAM sites to launch their missiles at him (at over 3000 ft/sec, which is faster than a 30-30 bullet) so their location could be determined and countermeasures taken. That was called "Dodging SAMs" but it was to protect the strike aircraft. One of these flights was again aiding a downed pilot and led to his third DFC.

Citation:

> "The President of the United States takes great pleasure in presenting a gold star in lieu of a third DISTINGUISHED FLYING CROSS to Lieutenant Wayne Paul Lotsberg for heroism and extraordinary achievement in aerial flight as the pilot of a jet aircraft attached to Attack Squadron FIFTY-SIX embarked in USS MIDWAY (CVA-41) On 17 May 1972. Lieutenant Lotsberg launched as a wingman in a section of aircraft assigned the task of surface-to-air missile suppression for a coordinated air wing strike into North Vietnam. After crossing the coast, Lieutenant Lotsberg's flight was diverted from the original flight plan to aid in the rescue of a downed pilot from another carrier. Despite intense anti-aircraft fire encountered in the rescue area, Lieutenant Lotsberg rolled in on the source and skillfully delivered his ordnance, scoring direct hits on an active shore battery. His section then came under attack from a surface-to-air missile site. In the face of five missiles, he aggressively launched anti-radiation missiles, silencing this threat. Despite continuous anti-aircraft opposition, he provided support until the rescue was successfully completed. The extraordinary aerial skill and devotion to duty demonstrated by Lieutenant Lotsberg reflected great credit upon himself and were in keeping with the highest traditions of the United States Naval Service."

Shot Down and Rescued

In November 1972, Wayne was shot down during a night reconnaissance mission and, thankfully, rescued due to it being night-time and lucky winds. Here is the account of his "water landing" and subsequent rescue in his own words:

About 1 am Nov 11, 1972 Ray Oswald (OZ) and I launched on a night "recce" (reconnaissance) flight over North Vietnam, looking for trucks moving equipment toward South Vietnam. I was the leader and I briefed on where we would coast-in and how to conduct our flight.

We crossed the coast and the weather was not very good with clouds and haze. We were south of Ha Tinh in Route Package 2. I was heading 240 and turning to about 210 and told OZ that I was going to head up the road at about 320 heading. At night we used a navaid and flew about 3 to 5 miles apart for safety since we could not see each other. We used the radio to help maintain flight integrity.

Following the road, I saw headlights heading south. I rolled into a dive with Rockeye weapon and delivered the weapon manually. During pullout I dropped a flare to light up the ground. I had previously dropped two flares for orientation to ensure where they were and only had one set of flares left. So, I told OZ to take a look and drop if he wanted to, which he did. He went down and I climbed up to roll in and drop my bombs when he called that the trucks were just south of the intersection, I had just dropped a flare over.

I was heading around the western side of the target at this time at about 10,500 feet. I changed my ordnance switches to set up a computerized bombing run.

LTV A-7 Corsair heading south, above the clouds

Well, then things started to become a can of worms. I had felt a tremendous grinding type vibration and my first thought was that I had forgotten to retract my Emergency Power Package (EPP), so I checked that and the handle was retracted.

Shortly thereafter I called OZ and said "I think I have a problem." About 15 or 20 seconds later I saw that PC2 (Power Control System 2 – main hydraulic source) gauge was fluctuating. It is supposed to be 3100 psi and had dropped to 2500, then back up, then down to zero.

I called OZ and said "I've got PC2 failure and am going to head for the water." I was turning the airplane toward the closest coastline and dropped my bombs when PC1 started fluctuating, which meant I was in big trouble.

I again called OZ that "PC1 is fluctuating and PC2 is at zero." He responded with "OK, get out your EPP and take it out of Control Aug" (a system setting that helps your control of the airplane). I pulled the EPP handle and it felt like it extended.

But the PC1 gauge still fluctuated a lot at this time. I was now heading about 120 which was not going to do me much good since that heading basically paralleled the coastline so I would be remaining over North Vietnam.

Within 30 or 45 seconds of the first vibration, both systems were at zero. At this time, I was heading about east, which was better, but still not heading directly for water, which was northeast. And now the flight controls were frozen. I could move the "stick" but the stick only tells the hydraulics what to do. So, I could do nothing to control the airplane flight. I had now resigned myself to the fact that I would have to get out of the airplane, it was just a matter of when.

When you have lost all hydraulic systems in the A-7 aircraft (which I was flying), as well as some other models such as the F-4 and F-105), the airplane will pitch full nose down, supposedly within 3 to 5 seconds. But my aircraft "froze" and was heading sort of east so I decided to stay with it until it did pitch down.

I called OZ and said "I'm heading east and still ok," but was just riding the airplane at that time. Then, lo and behold, the nose started to slowly pitch UP. I told that to OZ and he asked me if I had any trim available, which was a good call on his part. But trim can't work because it merely tells the hydraulic systems to reposition the control surfaces.

I was "riding it out" and the airplane wiggled a few times. When I lost both PC systems, I was about 9,000 feet. When the nose pitched up, I added full power thinking it was going toward the direction I wanted and stayed with it.

The nose had pitched up about 30 degrees (contrary to what the books and manufacturer claim) and in a slight left turn. This was miraculously taking me even more toward where I needed to go. The Good Lord must have kept the airplane in that climbing left turn for a little while.

Oz called me and asked my altitude and I told him I was at 15,200 feet. All of a sudden, the nose did pitch full nose down. At that time, I made the transmission "I gotta go, OZ" and out I went via ejection.

Northrop Grumman E-2C Hawkeye

(Listening to a tape that was recorded, the E-2 radar control airplane had a fix on my airplane at the time I ejected, so they knew my approximate location).

I really don't remember pulling the ejection seat handle but I remember holding onto the stick hoping it would do some good, but that was futile.

I pulled the alternate handle with my left hand and it seemed to be a while before the canopy slowly left the airplane and I felt air on my face. Then it seemed to be quite a while again before the seat rocket fired.

At the time I ejected, I was floating off the seat as the nose was pitching down and I remember worrying that I might hurt my back, which can happen if you are not sitting solidly on the seat when it fires.

The seat seemed to give me a very smooth acceleration up and I remember seeing the glow on the airplane from the fire of the seat rocket motor and then the canopy rails slowly going down as I rode up the ejection seat rails. I don't remember seat separation as such.

What I believe happened is that when I ejected, my brain kicked into "hyper drive" or something and perceived events at super speed. In other words, what took about one second was perceived as taking about 10 seconds. The events, canopy release, seat firing, riding up the rails out of the airplane, I saw in really, really slow motion when it took a fraction of a second.

About the time I realized that I was free and clear and "what do I do next?" I felt the parachute pop. I don't think I tumbled and the chute opened without any big jerk.

My concern was whether I was "feet wet" (over the water) or not. At the time all this started, we were about five or six miles inland over North Vietnam and heading more inland. I had resigned myself to the reality that I would have to eject whether or not I was over the water when the nose pitched nose down.

I was floating in my parachute and I was OK except I still did not know if I was over North Vietnam or the water due to the clouds and haze. I opened my "seat pan" which dropped and inflated my one-man raft. Then I inflated my flotation gear (life vest).

I floated down in my parachute for what seemed like a long while. Just after my parachute popped, I did get a glimpse of my airplane, which appeared to be about 3,000 or 4,000 feet below me in a steep nose down attitude.

I got out my survival radio and got the helmet cable. I connected the cable to my helmet but could not connect the cable to the radio, which would allow me to talk and hear with my helmet on. I remember being very frustrated that I could not get that attached.

I had a feeling that time was very precious. I turned my "beeper" (emergency radio signal) on and off a number of times to let others know that I was out of the airplane and that they should initiate a SAR (Search and Rescue) effort.

Then I was afraid I would lose my radio so I put it back in the pocket of my survival vest, still not knowing where or how high I was due to clouds and bad weather conditions. I was over clouds so I still did not know if I was "feet wet" but I felt that I possibly was over the water.

I could see Ray's airplane flying around a mile or two north of me. He popped a flare looking for me in the water. It turned out that he and other airplanes were calling me on the radio but I had stowed my radio and could not hear them. I was still floating down.

The flare that Ray had dropped was really nice because I could then look around and also see some sparkles below me through the haze so I knew that I was at least not directly over land. That also allowed me see that I was still fairly high over the water.

I floated down past the clouds that were off the beach and I prepared for water entry by getting ready to release my parachute fittings. I was remembering all the sea snakes I had seen when flying around the carrier and was wondering if they were any kind of a threat. Would they get me before rescue? Fortunately, I think they were all asleep.

I entered the water, and with the surprise (it is very difficult to judge height above water at night), I had forgotten to release my parachute for a second or two, but then I released it and it looked like it was deflating itself and sinking.

I was free of the parachute and saw my raft, found the lanyard attaching it to me, pulled it to me and climbed in. Getting in the raft was not a big problem (you pull yourself in on your belly). But in my haste to roll over and get my radio out, I fell out twice (they are small rafts) by being in a hurry, and because the wind and waves, which were 2 to 3 feet, added to the challenge.

"They are small rafts."

The third time I got in, I took my time, rolled over on my back, and felt pretty good about being in the raft. I was concerned for the helo pickup because the winds made the seas pretty rough, and the large, moving waves would make it difficult for the helicopter to hover, since the waves below it would be moving. Compounding this was the haziness, which made it even more difficult to maintain any visual flying reference to keep oriented.

I took my helmet off, got my radio back out and beeped a bit (I knew that the beeper was working because the radio would receive its own signal and I could hear it). Then I called OZ and told him I was in the water. After contacting OZ (it was really, really nice to hear his voice and to feel like I was no longer alone) I sat in the raft bouncing on the waves and turned on my strobe light.

OZ told me the helo was on its way. I felt very lonely in the water. I would get impatient waiting for someone to talk to me, just for the comfort of hearing a voice. I asked OZ how far off the coast I was and he gave me a vague answer, so I knew it was not very far. He just said "Hang in there" and "The helo is on its way."

I could see a lot of airplanes flying overhead. They were dropping flares and were trying to make visual contact with my strobe light, and therefore, me.

OZ told me to get my night flares. The helo wanted a flare so I popped one and they wanted another, so I popped another.

Bob Noll (sister squadron pilot – I knew their voices) had dropped a flare right over me, so they vectored the helo toward that flare and maybe 1/4 or 1/2 mile away they spotted my strobe and headed straight for me.

When the helo got close, I told them I was going off the air and was going to put my helmet back on. They rogered that and I laid my survival radio over the side because it was attached to me by a lanyard. I thought that if I needed it later, it would be there.

I got out of the raft and cleared it with no problem. If you stay in the raft, the downwash from the helo will blow you away from the helo and make pickup virtually impossible.

The helo dropped their swimmer into the water. He swam over to me and I assured him that I was OK. Then the helo moved in for the pickup. They were going to do a "D ring" pickup, attaching a connector to a D ring on my harness, rather than having me get into the sling or "Horse Collar".

With the haze and rough weather, the helo was having difficulty getting the sling to us and dragged it past us, but too far away for us to reach it. Finally, they dragged it close to us and just as they moved overhead, I grabbed the sling but they jerked it out of my hand and flew away. I was stunned. I asked the crewman with me "Where are they going?" and he answered "I don't know." (Remember I have mentioned the bad weather with the waves and haze. It turned out that just when they were ready for the pickup, the flare burned out, they lost their visual reference, experienced a severe case of vertigo, and had to depart to gather themselves to try again).

This really deflated my spirit and the crewman's, too, I'm sure. Rescue was so close at hand and now here we were abandoned. When they left, things really started to drag. They had dropped a flare about a mile south of us and they knew what was happening, but we did not. A feeling of abandonment tried to exert itself and it took great effort to maintain hope. Rescue seemed to be slipping away. I was worrying that they'd lose me after all.

To make matters worse, I went to retrieve my survival radio to find comfort in a friendly voice and to find out what was happening, but, like Murphy's Law, the radio lanyard was broken, so no communication. I have no idea how I lost the radio. I could see the helo, but they were not getting any closer.

The crewman pulled out his night flares (I was out of flares) and popped one that fizzled. With the rough seas, he had dipped it into the water just when it ignited so it went out. He was going to try another when I said "Wait a minute. Why don't you let me do that?" So, I took the night flare and popped it off.

I think it was about 20 minutes, which felt like eternity, before the helo started coming back toward us. They arrived overhead but it still took them a while to be able to drag the sling to us. (No, you are not able to swim over to it; that doesn't work.) All they could do, with the wind and waves, is estimate a path that would drag the sling to where we were, taking into account the nasty nighttime conditions. Finally, they did get the sling close enough so we could grab it, and they managed to hover long enough that the crewman could attach both of us to it. We got hoisted up to the helo and there was a bit of difficulty getting us in the door because the two of us were kind of tangled up.

Once inside the helo, I really, really felt relieved. It turned out that I had been about 4 miles off the coast. That close in daytime and I would have been picked up by a North Vietnamese fishing boat. Thankfully, it was about 2 am. I took out my survival water and drank it since I had swallowed about 300 gallons of seawater, never mind being stressed.

The crewman gave me a cigarette and we waited until we landed aboard the Truxtun, from where the helo had launched. The Executive Officer of the Truxtun was there to greet me.

I saw their ship's doctor who looked me over and decided that I was ok. The helo crew brought in a bottle of champagne which we smartly guzzled in a hurry and then the good doctor supplied a bottle of "Medicinal Brandy" to help calm us down.

I found out later that the helo crew should not have had to fly that rescue. They had not flown at night in 4 months, and had never made a night pickup before, not even for practice. AND that particular helicopter was "Down" (unavailable for flight for maintenance reasons) for loss of automatic stabilization control equipment and also for loss of Doppler Radar, which is a big aid for hovering anytime, and critical in rough weather over water. With all these challenges, the crew did an unbelievable job in rising to the occasion and completing my rescue. You see, no one is supposed to get shot down at 2 O'clock in the morning…

I stayed aboard the Truxtun until about 11 the next morning and got on a helo from my ship, the Midway, for a ride "home."

We landed and my commanding officer, Skipper Chatham was there and I intended to shake his hand, but I was so glad to be back, he got a big bear hug. Ray Oswald was there and he got a big bear hug. The Midway commanding officer, Captain Foley, was also there, but I did manage to contain myself and just shook his hand.

I went down to our flight surgeon and had 13 x-rays to see if I had broken anything and went to our squadron ready room and debriefed the whole thing.

Three days later, I was back flying over North Vietnam, which, like falling off of a bike, must be overcome by getting "back in the saddle." I ended up with 374 combat missions, 680 carrier landings, and one "water landing."

In Summary

To sum up, with 374 missions, many, many of them were DFC worthy, and there were numerous heroes flying against the same threats as I did. I lost roommates, fellow aviators and friends during this period. But I am grateful I had the opportunity to serve my country and am grateful to have done that which very few ever get to do; fly jets off carriers!

Artist Peter Chilelli's rendition of one of our flights: The aircraft were further apart than they appear in the painting, but Mr. Chilelli wanted to get everything into the picture.

With permission from Peter Chilelli and Fast Pixels. Painting titled "Welcome to Vietnam", based on actual event as described by Wayne Lotsberg: Flight lead, Wayne Lotsberg, left wing Keith LaFlair; Trail, Ed Beakley

Wayne Lotsberg's awards

I was invited to the Navy and Marine Corps Ball in Dallas in 1968 as the youngest, highest decorated Lieutenant in the Navy. The top left ribbon is the Distinguished Flying Cross award, with two gold stars for the second and third awards. The Distinguished Flying Cross is the highest award for flying heroism/valor. The next blue and gold award is the Air Medal showing 36 awards (for over 360 combat missions) and stars representing an additional eight awards for a total of nine individual Air Medals. The green and white is the Navy Achievement Medal with stars indicating a total of five awards.

Note: The Gold wings mean 'Qualified for flying off aircraft carriers and landing day and night and fully qualified instrument rating.' The Gold Wings idea is because only Navy, Marine and Coast Guard receive the same training, only they receive GOLD wings. Army and Air Force wings are silver.

Army and Air Force wings are silver. Gold wings means you have landed on a carrier!

And here is the young fella in about 1972. That is an A-7 Corsair II. This was taken at the VA-56 squadron home base in Lemoore, California.

Lieutenant Wayne Paul Lotsberg (1972), with LTV A-7 Corsair II

And in my later years

Wayne's daughter, Katie, created a web site with photos, tributes, & video clips of her father landing on an aircraft carrier. Readers are invited to visit:

https://kaitsdaily.wordpress.com/tag/vietnam-veteran/

Air War Over Southeast Asia

4th TFW F-105D Thunderchiefs at Takhli
Royal Thai Air Force Base, Thailand in 1965

Several of the pilots featured in this book flew out of Takhli: *Sawadee krap*

"Never tell the Platoon sergeant you have nothing to do."

22

Greg A. Nemeth
Staff Sergeant
United States Air Force

Vietnam

SSGT Greg Nemeth

My military adventure actually began with my dad, George. He served in the Army during WWII along with his four brothers. He was hit with a mortar on D-Day, June 6, 1944, and as a result he had so much shrapnel in his body that his upper body was speckled with black spots for the rest of his life. Even at age 82, sixty years after his service in WW II, a piece of shrapnel moved and severed an optic nerve in his left eye, blinding him.

During his time in France, his 5th Armor Division tank crew was captured and held captive in a farm house. One of the crew members was bayonetted through the stomach into a wall. My dad tried to help remove the bayonet but he was in turn bayoneted in the leg for trying to assist. In all the years following the war he never once mentioned a word about this incident or complained about it. Whenever he would get together with a couple of his Army buddies, my twin brother George and I, would sneak downstairs, and, sitting on the steps, would listen to their stories. Not until he was in rehab with a hip replacement shortly before he passed away, did I see his scars or have any idea what he went through. We owe a huge debt to those patriots who served in WW II.

My dad loved his Country and passed this love of country to his four sons; we all served in the military. George was in the Navy, Ed was in the Army, and both Gary and I were in the Air Force. Dad was my hero and continues to be an inspiration to me to this day.

This, then, is my story. I graduated from Perry High School, a small rural community located in Northeast Ohio, in June 1965. By October I had enlisted in the Air Force and was well into my training, eventually being trained as a C-130A Loadmaster. I am proud to say that of the 79 co-ed students in my high school senior class, a total of 38 people went into the military. The first in the class to join was my twin brother George.

My first airplane ride was from Cleveland Hopkins Airport to Lackland Air Force Base in San Antonio, Texas for basic training. From Lackland, I went to Shepard AFB to Loadmaster School. Little did I know the experiences that would unfold and the doors that would open by being a C-130 Loadmaster. My first assignment was at Lockbourne AFB in Columbus, Ohio where I was assigned to crew the C-130A.

Lockheed C-130A Hercules

Duties of A Loadmaster

The Loadmaster is responsible for making cargo placement decisions in the cargo area aft of the flight deck of the aircraft, actually loading the aircraft and managing the unloading process of whatever may be onboard. One of his most critical duties is to monitor the weight and balance of the aircraft since failure to perform this task accurately could lead to loss of aircraft control leading to mission failure. Maintaining the loading weight as close as possible over the center of gravity of the aircraft during take-off and landing is crucial. This is accomplished by centering the load and calculating the fuel burn which helps the pilot to maintain efficient control of the aircraft. Also critical is always being cognizant of any potential flight emergency where at any time the load may need to be jettisoned.

Lockheed C-130A Cargo Area

Once the aircraft is loaded and tied down with chains or straps, any passengers aboard are secured. Finally, the mission manifest or Form 365F must be completed, signed by the Pilot and Loadmaster, and submitted to Base Operations for approval before the flight can depart. The Form 365F contains the mission parameters such as destination, cargo load, passenger identification, fuel load, on board equipment, take off and projected landing conditions, and the total weight of passengers, crew, and baggage.

Next the Loadmaster sets up the power source for starting the engines using a ground turbine compressor or a ground power cart. On engine start he is outside the aircraft with headsets on to communicate with the pilot and also to post as a fireguard in case of an engine fire and to perform a final flight controls check of ailerons, elevator and rudder operation.

The C-130A, at the time was one of the largest most versatile combat planes in the inventory. It could be set up to accommodate a total of 92 fully equipped ground troops or 64 fully equipped paratroopers, or 74 litters to carry the wounded or any combination of the above. It could be configured for any type of air drop from large vehicles to small containers (small containers called A-22 containers are 52" x 54" at the base or larger containers 52" x 65"; it could drop 18 of the smaller containers at one time or 14 of the larger ones at a time.) A normal stick of 64 paratroopers could be dropped in about 1.5 minutes or it could perform a HALO (High Altitude Low Opening) jump, at maximum altitude.

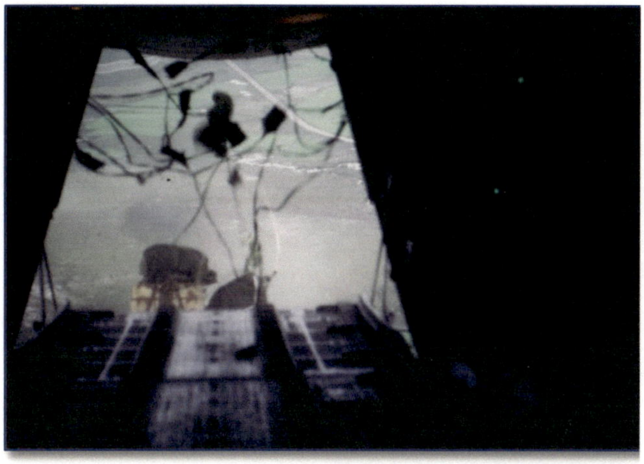

A-22 containers going out the ramp

A-22 containers with chutes open on the way to support the 26th Marines

The C-130A could deliver a cargo load by LAPES, (Low Altitude Parachute Extraction System), or a Ground Hook Extraction where a hook is dragged along the ground to snag a cable that would pull the load out of the aircraft during a low pass to exit the area. The two last mentioned are only for nonfragile, heavy loads like bunker timbers or fuel bladders. This technique limited our exposure to ground fire.

The adventure continued with orders to transfer to Mildenhall, England. We had not left the United States before we lost 3 of the 4 engines over Phillipsburg, Pennsylvania where we were fortunate enough to find a small airport to make an emergency landing. Since losing three engines was unusual, the entire C-130 fleet was grounded until the cause of the engine failures was determined.

Simultaneously with our arrival in England, Charles de Gaulle kicked the United States out of France. We were immediately reassigned to move all United States bases located in France to Spain, Germany and England. After Mildenhall, I was reassigned to Howard AFB in Panama where we trained in jungle warfare with the Army and shuttled back and forth to spend time at Fort Campbell with the 101st Airborne and Fort Bragg with the 82nd Airborne or at Fort Benning at the jump school. Back in the 60's they were still jumping out of planes before the transition to Air Cav. We would load 64 paratroopers, 32 in a stick, and make 4 or 5 drops a day.

After Panama the good times ended. I received my orders to Naha Okinawa and Vietnam. At this time, I should point out that it was just my luck to stay with the C-130A, the oldest 130 in the inventory, so if there was a mission where we might lose an airplane guess who was assigned the duty! Often my duties included "other duties as assigned" such as guard duty at Cam Rahn airbase.

My first crew assignment was with two captains that were at the end of their careers, and ready to retire. Being a young kid just out of high school they took me under their wings and showed me the ropes and how to make the best of the military. One of the captains, Captain Dick Sigman, would say, "I hope you paid attention to what you learned in tech school but just remember, the most important thing is to try not to get anyone injured or killed; don't tear up the equipment; and do what you can to complete the mission." To this day I use this motto whenever undertaking any large task.

On Guard Duty at Cam Rahn Bay

Most of my flying was out of Cam Rahn Bay. The C-130s were the largest combat planes in the inventory capable of short field landings, affectionately called the "trash haulers". We could and would haul the most payloads into small air strips and if there were a question about getting off the ground on short runways, we utilized JATO, (Jet Assisted Take Off,) bottles or rockets. Even with that we were big and slow which made us an easy target.

We also did forward air control (FAC), missions where we would fly at night. I would sit outside the 130 on the jump platform with a star scope, drop markers and 3 million candle power flares that would float down on parachutes turning night into day. I was looking for enemy troop movement. We then would call either Navy, Airforce or Marine fighters in to destroy the targets. Needless to say, we took a lot of enemy fire. We would return to base at the crack of dawn and the mechanics would start to patch up the aircraft so we could do it again the next night.

On the FAC Missions, the 8th Tactical Air Command fighter wing under Colonel Robin Olds was flying most of our fighter support. (Colonel Olds was later to head up the Air Force Academy in Colorado.) I was lucky to be invited to the Officers Club when one of the fighter pilots became an ACE; shooting down his 5th MiGs. What a thrill for a young airman. We did leaflet drops over Hanoi and Haiphong, dropping hundreds of thousands of leaflets while flying at the maximum altitude, wearing artic flight suits because of the freezing cold conditions, and sucking oxygen while kicking 180-pound boxes out of the aircraft. The joke was that if we were shot down by a SAM (surface to air missile,) which we would encounter often on these missions, the North Vietnamese would give us a stick with a nail on the end and we would have spent the rest of our lives picking up millions of leaflets!

I always realized my twin brother George had more wisdom then I. In Vietnam he was assigned to Naval Intelligence and Support at Camp Tinshaw in Danang, the largest forward operating base in Vietnam which supported the Marines and Army in missions going north to the DMZ. To this day I have never been able to figure out how one of my Air Force co-pilots, who was also an ex-Navy commander, knew I had a twin brother in Vietnam. By the way, it was against regulations to have two family members in the war zone at the same time. Instead of sending one of us home, he would call George's commander and have George sitting in a jeep in full combat gear to meet our plane on our supply missions to Khe Sanh. For the 26th Marine Regiment, we were the main source for survival. We would fly up a valley to the drop zone below the mountain tops taking a lot of enemy fire. When we would get to the drop zone, we would jettison 30,000 pounds of supplies in about 20 seconds. The plane would then shoot straight up. When the dust would clear we would secure the plane and head back to the base at Danang to drop George off. Where does the wisdom show up? George, still shaking, would say being in the Navy if the ship went down, I could swim a long way, but if your plane went down, I couldn't fly very far! George willingly would go on to fly several more missions with us. In recognition of these missions I received an Air Metal and a Marine Presidential Unit Citation but since George was not supposed to be on the aircraft, he unfortunately received no recognition. George passed away shortly after our 71st birthday of complications from Agent Orange.

George Nemeth 1947 -2018. RIP. He served with courage and honor.

My two DFCs were awarded for action in support of the 26 Marine Regiment while they were surrounded by the North Vietnamese and Viet Cong at Khe Sanh in 1968. All the routes to resupply the area were cut off or too dangerous to accommodate vehicles. Consequently, they would repeatedly run out of food, water, medical supplies and ammunition. Our resupply missions were critical to their survival. At first, we would land and off load on the ground. On the very first mission we took ground fire when we landed and pulled into the unload area. As soon as I opened the ramp and door a rocket or mortar hit behind the plane and showered us with shrapnel. Soon after, it was determined that the max time we could spend on the ground was 8 minutes which was impossible to complete the task. So, we started dropping containers by parachutes or we would do LAPES or Ground Hook Extractions where we would extract the heavy loads out of the plane as we were taxing down the runway on our way out. The only problem was the constant ground fire we would take going in and out. The route going in was along a valley and at times we were the optimum target of which the Northern Vietnamese often took advantage. And on final approach we were lower than the mountain tops so we took fire from above. On one such mission I was sitting on a load of ammo for the 105 and 155 guns thinking that the safest place on the plane was in the wheel well where the landing gear was stowed while in flight, thinking that an enemy round wouldn't go through all that steel. Soon after, a round did go through that area blowing out the air/heat duct above my head. From that point on I realized that when your number is up, it's up so don't worry about it.

LAPES load ready to go

To perform my duties as a Loadmaster I had to move around fast to get the load ready to drop, open the ramp and door, and extract the load and button up the back of the plane. One of my options was to fly with a parachute which is heavy and bulky. Plus, on the way down the G force would not let me get to a door, open it and jump in time for the chute to deploy so that was not a good option. My next option was the Marine flak vest which had several steel plates positioned to protect your vital organs. As I found out earlier on a mission where the round came through the wheel well, if it hit the steel plate in the flak vest it would shove through my body not stopping at the steel plate in my flak vest. I unfortunately realized my best option was to fly in an olive drab tee shirt.

My first Distinguished Flying Cross was awarded after the actions of March 6th, 1968. While flying a tactical emergency resupply air drop of vitally needed supplies to the surrounded ground forces at Khe Sanh we came under intense ground fire with bullets flying through the fuselage. I was able to set up and complete the drop. When we returned to home base the maintenance crew put green duct tape on the bullet holes in the C-130. I counted over 40. Talk about luck, the flight crew came out unharmed, but most important, we successfully delivered the needed supplies.

THE UNITED STATES OF AMERICA

TO ALL WHO SHALL SEE THESE PRESENTS, GREETING:

THIS IS TO CERTIFY THAT
THE PRESIDENT OF THE UNITED STATES OF AMERICA
AUTHORIZED BY ACT OF CONGRESS JULY 2, 1926
HAS AWARDED

THE DISTINGUISHED FLYING CROSS
(FIRST OAK LEAF CLUSTER)

TO

STAFF SERGEANT GREGORY A. NEMETH

FOR
HEROISM
WHILE PARTICIPATING IN AERIAL FLIGHT
6 March 1968

GIVEN UNDER MY HAND IN THE CITY OF WASHINGTON
THIS 27th DAY OF December 1969

GEORGE S. BROWN, General, USAF
Commander, Seventh Air Force

SECRETARY OF THE AIR FORCE

DFC Certificate 6 March 1968

On the drop missions we would try to fly with two Loadmasters. At the time, my boss, Mel Hugging, a Technical Sergeant, was the NCOIC, the non-commissioned officer in charge. Consequently, because I never said no to any assignment, I was the appointed the assistant NCOIC. This title that meant when no one else was available, I flew regular missions and emergency resupply missions by myself. As a result of one of those missions, I earned my second DFC on April 5, 1968.

While attempting an aerial resupply mission to Khe Sanh with critically needed supplies, we came under intense automatic weapons fire resulting in heavy damage to the fuselage and fuel tanks. With leaking fuel and bullets coming thru the floor, we pressed on to complete the drop. We had been previously briefed on the need of our supplies and knew that without them the Marines could not defend themselves. It was always the same story - our forces at Khe Sanh would need to be resupplied, we would load up and fly in where we always took ground fire and just prayed that we would return to base in one piece. When we did, it was such a good feeling knowing we aided in the survival of our ground troops.

CITATION TO ACCOMPANY THE AWARD OF

THE DISTINGUISHED FLYING CROSS

TO

GREGORY A. NEMETH

Sergeant Gregory A. Nemeth distinguished himself by extraordinary achievement while participating in aerial flight as a C-130 Loadmaster at Khe Sanh, Republic of Vietnam, on 5 April 1968. On that date, Sergeant Nemeth flew on a tactical emergency resupply airdrop of vitally needed supplies to friendly combat ground forces besieged at Khe Sanh. Under extremely hazardous and challenging conditions, Sergeant Nemeth aggressively pursued and successfully completed this crucial mission. The professional competence, aerial skill, and devotion to duty displayed by Sergeant Nemeth reflect great credit upon himself and the United States Air Force.

DFC Citation, 5 April 1968

DFC Award

After serving two tours in Vietnam and numerous other assignments I was working on my 3rd Lockheed 1,000-hour pin when I was honorably discharged.

Thank you to the United States Air Force. You provided me with quite an adventure. I have always considered myself one of the lucky ones; counting my blessings that I was able to return home unharmed and able to return to a normal, everyday life and finally, thank you to The Distinguished Flying Cross Society, Villages Chapter, for giving me the opportunity to tell and share this story.

SSGT Greg A. Nemeth

23

Robert (Bob) S. Sherman
Captain
United States Air Force

Vietnam War

Young Electronic Warfare Officer Bob Sherman

The Douglas B-66 Destroyer

The Douglas B-66 Destroyer - aka - Lead Sled. Deliveries to the USAF began in 1956, with 145 of this model produced. The RB-66C was a specialized electronic reconnaissance and Electronic Counter Measure (ECM) aircraft with an expanded crew of seven, including the additional electronics warfare experts. A total of 36 of these aircraft were built with the additional crewmembers housed in what was the camera/bomb bay of other variants.

In 1966, these planes were renamed into EB-66C. The unarmed EB-66B/C/E aircraft flew numerous missions during the Vietnam War. They not only helped to gather electronic intelligence about the North Vietnamese defenses, but also provided ECM protection for the daily bombing missions of the F-105s by jamming North Vietnamese radar systems.

The EB-66B/C/E was removed from USAF service by 1975 and most examples either scrapped in place or placed in storage for eventual scrapping.

Bob Sherman in position as EWO

I was an electronic warfare officer (EWO) and it was my job to detect and jam enemy radar, thus preventing the tracking of US armed forces on combat sorties. I was stationed at Takhli Royal Thai Air Force Base, 140 miles north of Bangkok, Thailand from May '67 thru May '68 and flew 121 combat sorties.

B-66 fiery crash scene, August 28, 1972

Pictured above is a fiery crash scene from my last flight. On August 28, 1972, while on a training mission out of Spangdahlem AFB, Germany, our aircraft experienced mechanical problems and our pilot was forced to do a high-speed abort while on take-off (160 knots with 27,000 pounds of fuel).

We all lived. The B-66 did not fly again.

Distinguished Flying Cross Awarded for actions during the Vietnam War

The first DFC was awarded for a routine combat sortie off the east coast of North Vietnam, near Haiphong harbor located in the Red River delta. Our mission was to provide ECM support for the USAF/USN attack forces. While on station, a Navy aircraft was shot down and the pilot successfully ejected. He landed in the waters about 20 miles off the coast and you could hear his voice as both the North Vietnam and US Navy forces carried on the search and rescue efforts. We remained on station until the Navy successfully recovered the downed officer.

Citation:

> "The President of the United States of America, authorized by Act of Congress, July 2, 1926, takes pleasure in presenting the Distinguished Flying Cross to First Lieutenant Robert S. Sherman, United States Air Force, for extraordinary achievement while participating in aerial flight as an EB-66C Electronic Warfare Officer over North Vietnam on 24 October 1967. On that date, Lieutenant Sherman's unarmed electronic warfare aircraft provided valuable electronic

support against the unfriendly radars which were tracking the combat operations flown against Phuc Yen Airfield. His exceptionally commendable performance in the face of personal danger from hostile fire was a vital factor in denying electronic information to the unfriendly foreign forces and a major contribution to the successful outcome of this dangerous tactical mission of the United States Air Force. The professional competence, aerial skill, and devotion to duty displayed by Lieutenant Sherman reflect great credit upon himself and the United States Air Force."

The second DFC was awarded again for a routine sortie to protect the US air operations at the western edge of North Vietnam. On most sorties, our aircraft would establish an oval orbit to protect the attack force and we had just arrived on station. That day I notified the pilot there was unusually heavy surface-to-air missile and anti-aircraft radar activity. Later during the sortie, we had an enemy aircraft radar lock-on and our aircraft was attacked by a MiG-17. Since all the EB-66C had for defense was a chaff dispenser and ECM equipment, the pilot took evasive actions, we alerted our Close Air Patrol (CAP) support aircraft and successfully returned to Takhli. That afternoon, another EB-66C flying the same mission was attacked and shot down. The surviving crewmembers spent the rest of the war as POWs.

Citation:

"The President of the United States of America, authorized by Act of Congress, July 2, 1926, takes pleasure in presenting a Bronze Oak Leaf Cluster in lieu of a Second Award of the Distinguished Flying Cross to First Lieutenant Robert S. Sherman, United States Air Force, for extraordinary achievement while participating in aerial flight as an EB-66 Electronic Warfare Officer in Southeast Asia on 19 November 1967. On that date, Lieutenant Sherman participated in an electronic countermeasures support mission against radar directed anti-aircraft weapons and surface-to-air missiles which threatened a large combat force. His outstanding determination and professional performance in the face of immediate danger from hostile fire was a significant factor in the overall success of this vital mission. The professional competence, aerial skill and devotion to duty displayed by Lieutenant Sherman reflect great credit upon himself and the United States Air Force."

24

Gary M. Spooner
Chief Warrant Officer
United States Army

Vietnam War

Gary M. Spooner, May 11, 1970

I was born in Manchester, NH on Sept 11, 1947. Never realizing that my birthday would be such a sad and historic day. Since my Dad was being transferred around the country for his job, we lived in Hookset, NH, Sugar Hill, NH, Palm Springs, CA, Pembrook, NH and St Johnsbury, VT. I graduated from St Johnsbury Academy, an excellent private school, because the town did not have a public High School—It is still like that to this day. While in High School I lettered in Baseball, Football, Track and the Ski Team. I was the Captain of the Ski Team. After Graduation, I moved on to the University of Vermont in Burlington, VT.

While at the University of Vermont in my second semester Junior year in late January 1968, I decided to skip the remainder of the Semester and become a Ski Bum in Stowe, Vermont. I drove up to Stowe and the first day found a job at The Toll House Inn, a Luxury Inn, at the top of the Mountain Road adjacent to the Ski Area of Mount Mansfield. The job was being a Bell Hop six days a week, working from 6:00 AM till 3:00 PM three days a week and working from 3:00 PM to 11:00 PM three days a week. That meant that I could ski four days a week and the job included a Season Pass. The plan was to go back to school at the University of Vermont in the Fall Semester. I was an avid skier and a Ski Racer and had some experience as a Ski Instructor at Bolton Valley Ski Area in Bolton, VT. The Ski School Director was Art Furrer, who was my Race Training Coach when I was on the Franconia Ski Team at Cannon Mountain, NH.

Three weeks into this fun-loving fantasy, I received a call from my Dad, that I had received an Invitation from the US Government to be Drafted into the US ARMY. What am I going to do now? I traveled to Burlington, VT to visit all the Recruiters. I decided that I wanted to be a Pilot—Air Force: required a college degree—Navy: required a college degree—US Army: Warrant Officer Helicopter Flight Training Program only required a High School Degree. I was on that one. Had to take two days of testing to qualify for Flight Aptitude. I knew I could do that. Also was able to get an Extended Entry Deal that allowed me to not join the Army until May of 1968. That meant that I could ski the rest of the ski season.

I entered the US Army Warrant Officer Helicopter Flight Training Program in late May of 1968. Traveled to Fort Polk, Louisiana for eight weeks of Basic Training. Fort Polk in the heart of the summer is just a beautiful place and very humid and hot. Late July, I traveled to Fort Wolters, Texas to begin Flight Training. First four weeks was Pre-Flight Training to see if you had the guts to complete the almost year long Flight Training. About a third of us got rejected from the Program for physical reasons, lack of flight skills or for failing flight exams. At Fort Wolters, I soloed in the TH-55 made by Hughes Aircraft.

TH-55 Training helicopter

Next Journey was to move to Advanced Flight Training at Fort Rucker, Alabama. Forgot to mention that during my Christmas Leave back to Vermont, I married a lovely Canadian lady from Montreal by the name of Johanne Cheverton. I had dated her prior to going into the military, so we decided to tie the knot after I was in flight school. She traveled with me to Fort Rucker and we rented a 10 X 50 Mobile Home in Daleville, AL.

After graduating from Flight School, we had a 30-day leave, then off to Report to Vietnam, I arrived in Vietnam on June 20, 1969. I was assigned to C Troop, 7th Squadron, 17th Air Cavalry Regiment at Lane Army Heliport about 10 miles from Qui Nhon. My roommate from Flight school, John Wilcox and I followed the same path to Vietnam and ended up in the same unit. When we reported to our Troop Commander, Major Joe Tobin, His first words were, "You and You are going to Cobra AH-1G flight training school in Vung Tau next week." So off to Vung Tau we went for two weeks to be Qualified in the Cobra Gunship. The in-country transition in Vung Tao was two weeks in duration, one week to learn how to fly the aircraft and one week to learn to fire the armament systems. This same transition in the U.S. at Hunter Army Airfield was six weeks. In Vietnam 'you learn on the run'! When you get back to your unit you are still learning.

The Mission of the Air Cavalry is to search out and destroy the enemy from the air. We used two LOH's (Light Observation Helicopter) OH-6A's called Cayuse. These two LOH's would fly over the tree tops in areas that we were searching for the enemy looking for troops, Hooches (Huts that the enemy would use to sleep in), caves and anything that we could find like trails to get movement from the enemy. Our LOH's would fly in formation with the Lead LOH looking for the enemy and the Trail LOH keeping right behind the Lead to lay down a burst of gunfire if the Lead LOH got shot at. This happened quite frequently, and when the LOH would take fire from the enemy on the ground, the observer flying with the Pilot of the LOH would throw out a smoke grenade marking the spot. High Above the LOH's were two AH-1G Cobra Gunships circling to cover the LOH's. As soon as the LOH called "Taking Fire" and the smoke came up through the jungle the Lead Cobra would roll in from about 1000' AGL (Above Ground Level) and fire onto the smoke with bursts of Mini-Gun and 40 MM grenades The Aircraft Commander flying in the back seat of the Cobra would fire 2.75 inch rockets from his wing stores and lay down as much lead as he could. The Cobras would make two or three passes with Lead going first and Trail following. We would go back to 1000' AGL and let things cool down for a bit and then the LOH's would go back in to observe what damage we did or if we were not lucky. This was a bit tense for the LOH Pilots. As this is all happening our C&C Ship would be flying about 1500' AGL observing all that is happening and directing the operation. The Command and Control ship would be flown by our Troop Commander with another Pilot doing most of the flying.

Just to give you an idea of what the AH-1G Cobra capabilities are, I will list its great Fire Power.

"Chariots of Fire" by Joe Kline, Vietnam War artist, showing C Troop 7/17 Gunships

I flew Cobra AH-1G Gunships for just over 6 months and rode in the front seat as the Co-Pilot/ Gunner. The front seater fired the Turret which consisted of a Mini-Gun (A six-barrel rotating Gatling gun) that fired link belted 7.62 rounds at a rate of 2,000 rounds per minute or 4,000 rounds per minute for four to five second bursts. The gun would over heat after that and the cut off switch would stop the mini-gun from firing. The other weapon on the Turret was the 40 MM Grenade Launcher that fired 40 MM grenades at a rate of 300 rounds per minute. You could fire bursts of 5-7 seconds and lay down a lot of explosives. The Aircraft Commander who was the pilot in the back seat had control of the wing stores that held the four Rocket Pods. We had a variety of combinations of seven shot pods and 19 shot pods. Basically, we had 17-pound Rockets in the seven shot pods and 10-pound Rockets in the 19 shot pods. We could carry four each 19 shot pods, that would hold a total of 76 rockets, but we were confined to only using 10-pound Rockets for weight considerations. We carried either HE (High Explosive) Rockets or White Phosphorous Rockets that were good for burning Hooches down. We also could carry Flechette Round Rockets that would fire "Nails" (one and half-inch long nails with little fins on them), in half a football field we could land one nail in every square foot of that area. The Rocket would explode with a Puff of Red Smoke at about 60 feet above ground level. This rocket was only used in open areas, as trees would interfere with the Nails. When firing rockets, you used either your Outboard Pods or your Inboard Pods, depending what rockets you wanted to fire. Of Course, that while keeping the diving Attack Helicopter in flight trim. When rockets were fired, they fired off in pairs, one from the left side and one from the right side. Our Selector Switch allowed us to fire one, three, five or seven pairs. When the trigger was pulled, a pair at a time would go off, with about a half second delay between each pair. Firing three pairs might take three seconds.

In late January 1970, I got antsy about flying front seat and requested a switch to the Slick Platoon flying HUEY'S, the UH-1H's that we used to fly troops into remote Landing Zones (LZ's). That consisted of inserting our Blues (Ground Troops) into a four aircraft formation into landing areas that probably never had a helicopter land there before. Most of our LZ's were areas covered with high elephant grass and rarely did we get into an area that would allow a perfect level landing. Most insertions were done at a 2- to 6-foot hover and those troops jumped very quickly, with minimum time near the ground. When we inserted our Blues, we carried six troops on each ship and inserted 24 grunts (Infantry) on the ground. Of course, when our ground troops finished their ground mission, we would come back and pick them up, usually out of a different LZ than where we inserted them. We called that the Pick Up Zone (PZ).

Let me tell you about the Happiest Day of my year in Vietnam (excluding the day I left). While still stationed at Lane Army Heliport near the Village of An Son, I was waiting to hear about the birth of my first child with my wife, Johanne, living in South Burlington, VT. On March 13, 1970 at about 0630 (6:30 AM) we were just taking off after first light from Lane Army Heliport to fly to our Staging Area at LZ Two Bits at Bong Son near LZ English. We had just left the traffic pattern of Lane Army Heliport and my Troop Commander came on our Fox Mike company radio and called, "Blue 36 (my call sign), this is Yellow Scarf 6, come in." I responded, "Yellow Scarf 6, this is Blue 36, go ahead. "Major Jim Hahn (Yellow Scarf 6) said this: "Last night we received a Telegraph from the Red Cross, that says 'We are happy to report that your wife has had a new child on March 11 at 7:11 AM in Burlington, VT and Mother and Child are doing fine. The new child happens to be a Baby—' Then he paused for about 15 seconds—'A Bouncing Baby Boy by the name of Jay Mark Spooner.'" Right then I asked my co-pilot to take the controls, as this was a bit of an emotional experience for me. Of course, everyone in my Troop who was flying and, on the radio, heard this. I got a whole bunch of congrats over the radio from all of my fellow pilots. When we landed at LZ Two Bits, everyone came over to me and congratulated me with my new son being born. Of Course, in those days, it took a day and a half to get a telegraph and we were on the other side of the International Time Zone, so we were always a half a day ahead. My Dad had called the Red Cross the morning of March 11th that my son was born, so they sent it along to me in Vietnam. In those days, the telegraph was our best line of communication. So that day March 13, became The Happiest Day of my year in The Republic of Vietnam.

In late March of 1970, C Troop moved to the beautiful and lovely Ahn Khe Air Field, affably called the Ahn Khe Golf Course. (There was no golf course there.) We set up our new operation there and still operated in most of the same territory that we had already worked in, mostly the area North of Phu Cat AFB including LZ English, LZ Uplift, LZ Two Bits and the An Lo Valley.

Gary, March 1970 in An Khe, an Aircraft commander at age 22

Here's where the story turns to another page in the history book of Vietnam US Army Air Cavalry fighting to save a country from the Communist North Vietnam. The first week of May, 1970 was the First Legally Announced Invasion into Cambodia. Read between the lines. (We had been there before). The Ho Chi Min Trail was the road that the North Vietnamese used to haul equipment, personnel, weapons, ammunition and supplies from North Vietnam to South Vietnam; situated along the Cambodian Border. We worked areas west of the Ho Chi Min Trail and even found two Soviet tanks traveling south on the Trail. We didn't want to atack them with Cobra Gunships, so we called in the Air Force with two F-4 fighters. They blew those two tanks up lethally by dropping 500 lb bombs on them.

While working that area for about two weeks we were staging out of a Montagnard village at LZ Du Co just east of the Ho Chi Min Trail by about two miles. It was advantageous for us to use the Fire Support Base there to re-fuel our Aircraft and re-arm our Cobras, Hueys and LOH's.

Now, getting to my story of receiving a Distinguished Flying Cross, which ranks fourth in order of precedence as a valor decoration for the Armed Services.

On May 11, 1970, we were working an area about 15 miles into Cambodia near the Triad area; that was in the area where South Vietnam, Cambodia and Laos joined together. While cruising over the tree tops our LOH Platoon Leader Captain John Sartor (Red 6), saw some North Vietnamese Army (NVA) troops on the ground and immediately started taking fire. As John Sartor was taking fire, he got shot in his flight helmet. He thought his head had exploded, but realized he was still alive. He lost control of the aircraft at about 25 feet above the ground and made a crash landing into the ground. John had an observer in the right rear seat and an observer in his left front seat. When they crashed John and his rear seat observer exited the aircraft and ran to take cover away from the LOH as it might explode or catch fire. The left front seat observer exited the aircraft and ran the opposite direction into the jungle and hid as far away from the aircraft to feel he was safe; he was hiding in a treed area with trees about 20 feet high.

After this crash the Cobras came rolling in and making passes over the area firing mini-gun, 40 MM grenades and rockets. They made passes in two directions so as to wipe out the NVA troops on the ground as quickly as they could. Because of the fire fight that was going on, our C&C Ship flown by Captain Peter Kacerguis (Yellow Scarf 6) and Warrant Officer Mike Sipsey called in the Air Force to bring in some F-4 Fast Movers with 500-pound bombs aboard. It took them about 20-30 minutes to arrive on station. In the mean time, CPT Kacerguis called in our other two Cobras on stand-by to relieve the two Cobras as they were getting short on fuel and had run out of ammo and rockets.

Two fresh Cobras arrived on station and began to lay down more rockets. When the two F-4's arrived, they started dropping 500 lb bombs. CPT Kacerguis spots where CPT Sartor and his observer were hiding and noticed there was a small area where CPT Kacerguis could land and pick up CPT Sartor and his observer. This was done successfully. CPT Sartor got shot only in his flight helmet. He had a round go through the right rear side of his helmet and the round skinned about three inches of his skull and exited his helmet in the top left side of his helmet. Fortunately, CPT Sartor was OK. CPT Kacerguis called for replacements of the two F-4's on station to bring in two more. About that time the other observer who was hiding in the treed area got on his PRC-25 emergency radio and got a transmission through to CPT Kacerguis in the C&C ship.

They got a spot on his location which was approximately 150 meters from where CPT Sartor was on the ground. I, (Gary Spooner,) got a radio call from CPT Kacerguis on our Fox Mike (FM) radio like this...I was the Aircraft Commander of UH-1H Ser. # 67-17328. "Blue 36, (my call sign), this is Yellow Scarf 6 come in." I answered, "Yellow Scarf 6, this is Blue 36 go ahead." 6 responds, "Blue 36. Do you have a rope ladder on board?" I answer, "Yes Sir, we do." 6 responds,"36-Get out here as fast as you can because we have an Observer to pick up off the ground." I answer, "We are cranking now and on our way." I cranked my Huey, a UH-1H Ser. # 67-17328 (I like to repeat that number of my Old Girl) as quickly as I could. We had already flown about four hours that day and already had made two insertions and extractions of our Blue Troops. We were staging at LZ Du Co with our Fox Mike Radio on listening to what was happening about 15 miles away.

After getting airborne I called Yellow Scarf 6 and reported that I should arrive in about 7-8 minutes. Yellow Scarf said to me, "Report back when you are about one minute out, as we will drop a smoke from one of the LOH's in the area where the observer is hiding and make your approach to the smoke at tree top level facing east." When I reported in close to the operation, I could see the Cobras starting to make their passes to keep the NVA heads down, in case any of them were still able to fire back. I also saw the two F-4's starting to make their passes beginning to drop their 500 lb bombs.

Within 45 seconds or so, I had arrived on station making a fast low-level approach to the smoke coming up through the trees. CPT Kacerguis was guiding me in over the tree tops, to where we would come to a hover and threw out our rope ladder on our left side. I was flying in the left seat so I could look out my door window and see the observer running to our aircraft and the rope ladder that was 25 feet long like a sitting duck in a fish-barrel shooting contest. As I was on short final, I could hear automatic weapons firing that were the NVA's AK-47's. I was also seeing the Cobra gunships making their runs and then heard the F-4's dropping their 500-pound bombs. This was a really loud war going on.

As I came to a hover, my crew chief on the left side said to me he could see our observer running to the pick-up area. Of course, both my Crew Chief and my Door Gunner were firing their M-60 machine guns to distract any enemy on the ground. I hovered there for at least 2 to 2-½ minutes waiting for our observer to climb that rope ladder. The observer still had his M-60 machine gun slung over his back and still had on his chicken plate. ...All I could think of was this is going to be my last day and was I going to get shot down any second. That was the longest 2 ½ minutes of my life, waiting for the observer to climb in to my aircraft. When he was almost to the skid area of the Huey, my crew chief said "He is almost in.", and my Co-Pilot, who was watching over his left shoulder said, "He is in. Let's go!" I pushed that stick forward and nosed over that helicopter and blasted out of that area to the southeast as fast as that aircraft would allow pulling as much pitch (Collective) as I could. We flew tree top level for about 2000 meters until I got the ship up to a 100 knots (115 MPH) and then I pulled back the stick as we climbed into the sky at 2500 feet per minute with full collective pulled. After we reached around 1500 AGL we leveled off and flew back to LZ Du Co. BOY! WAS I RELIEVED! I was a Chief Warrant Officer II at 22 years of age and just survived what could have been my last day in Cambodia of the so called Vietnam War.

"God's Own Lunatics" by Joe Kline personalized by the Artist with my aircraft #67-17328 and C Trp/ 7th Squadron/17th Air Cav markings.

Even though the NVA that were still alive on the ground were firing away, we never took a round and got out of there unscathed with no bullet holes in the aircraft or any of our crew.

A few days later, while we were still on the Cambodian Operation, flying back to An Khe from LZ Du Co, we were flying near Pleiku and my engine chip detector light came on. From my old girl #17328. Thanks for the warning! We were very close to our Squadron Maintenance at Camp Enari, just outside of Pleiku. I called Camp Enari Tower and reported an emergency and was landing at the 7th/17th Maintenance area. I did an Autorotative Landing with power at the bottom and hovered over to the Maintenance Quonset Hut. I shut down the aircraft and an old gnarly Specialist 6 came out who was our Maintenance Specialist. We pulled out the magnetic engine chip detector plug on the side of the engine. The Specialist felt for the engine oil and metal chips on the plug and said to me, "Sir, those metal chips are within tolerance and you should be fine." I questioned him on that and he said he was right. OK, got to go along with what he says. OK. We fire up the ship and the Eng Chip Detector light is off. We take off out of Camp Enari with four Crew on board (A/C, Peter Pilot, Crew

Chief, Door Gunner) and six Troops on board for a total of 10 personnel. We climb out of the traffic pattern and the old Engine Chip Detector Light comes on again. Holy Crap! This is not good. We whirl around a 180 degree turn and I get on the radio to Camp Enari Tower, that I am coming back for an emergency landing for the second time. We do an Autorotative Landing with power at the bottom. We are hovering over to the Quonset Hut at a three-foot hover and, lo and behold, the 1400 Shaft Horsepower JET TURBINE ENGINE QUITS and we drop her to the ground with a Hovering Autorotation. As soon as I get the rotors stopped, I jump out and run over to that Specialist 6 and proceeded to chew him out and hollering to him "You could have killed us all!" Thank God, all 10 of us survived that one.

Got home to Vermont on June 20, 1970 at 11:30 PM (one year to the day) and was greeted by my wife, Johanne, my mom, my dad, my sister and guess what, a 3 ½ month old baby boy by the name of Jay Mark Spooner. Was I ever happy to see my new wife and my beautiful son, who was wearing a red, white and blue jump suit. What a nice welcome home. I had finally made it home!

My last bit of DFC information: When leaving South Vietnam, I never realized that I had been put in for a Distinguished Flying Cross. After Vietnam, I was assigned to Germany and was chosen to be VIP Air Taxi Pilot for the 56th Aviation Detachment in Coleman Barracks, Germany, just outside of Mannheim. We supported TASCOM, (Theater Army Support Command-Europe) and provided both helicopter and fixed wing VIP flight services to the command. Our passengers were full Colonels, Generals and Dignitaries. While there I flew three Star Lieutenant General Charles Eifler, many 1- and 2-Star Generals, whose names I have forgotten, Col George Patton Jr and most notably The Secretary of the Army, Stanley R. Resor, who signed my Distinguished Flying Cross Certificate. I should have given him a resounding Thank You, but we spoke only business to our guests. After being in this unit for about a month in late September, 1970, my Company Commander, Colonel Butler called me up and asked me if I were going to attend the formation at noon that day. My comment was, "Sir, I wasn't planning to be there." Col Butler said, "I need you to be there." Of course, I responded with a quick "Yes Sir, I will be there." When the Formation was called together, Col Butler announced, "CW2 Spooner Front and Center." This rather shocked me. "What is this all about?"

Col Butler announced that I was receiving the Distinguished Flying Cross for Heroism while participating in Aerial Flight in The Republic of Vietnam. Actually, it was in Cambodia!

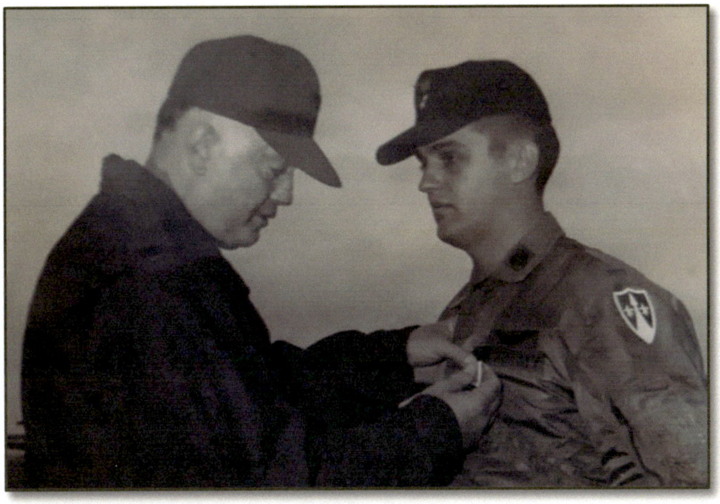

Receiving the DFC

Gary Spooner Today

"That is my Story and I am Sticking To It!"

P.S. – In 1976 I smartened up and graduated from New Hampshire College (now Southern New Hampshire University) with a Bachelor's degree in Business Administration.

25

John H. Webb
1st Lieutenant
United States Army

Vietnam War

Young John H. Webb

I was born in December 1946, in Valdosta, Georgia, among the first of the Baby Boomers. I graduated from Dixie Hollins High School in Saint Petersburg, Florida in 1965 and joined the U.S. Army in September of the same year, delaying my college education.

I was initially assigned to the Army Security Agency (ASA) and later graduated from Artillery Officers Candidate School (OCS) in March 1967, becoming a Commissioned Officer. I was married immediately after graduation. I then attended Helicopter Flight School, graduating in April 1968, immediately followed by a one-way ticket to the Vietnam War in late May 1968.

The Communist Tet Offensive of 1968 had occurred a few months before while I was still in flight school. Rounding up and eliminating the remaining remnants of the defeated North Vietnamese Army and Vietcong forces that had attacked South Vietnam was in full swing.

I joined the "Little Bears", A Company, 25th Aviation Battalion, 25th Infantry Division at Cu Chi base camp, twenty miles north of Saigon, Republic of South Vietnam. The third leg of the Ho Chi Minh Trail dumped into South Vietnam through the porous Cambodian border just thirty miles northwest of Cu Chi, supplying five Vietcong battalions in the Cu Chi District. The Vietcong were well armed and aggressive throughout the area. The Little Bears flew UH-1D "Huey" helicopters providing general and combat support to the 25th Infantry Division in its mission to protect Saigon, the capitol of South Vietnam. The years of 1968 and 1969 were the height of America's troop commitment to the Vietnam War and would prove to be very challenging years.

1st DFC: 8 October 1968, Cu Chi, RVN Night Emergency Re-supply Missions

Out of all the emergency resupply missions that I flew in 1968-69, one of them was particularly noteworthy. I was on standby that night and at about 9:30 PM, I was told by Flight Operations that we had a re-supply mission. We were to pick up an Artillery Forward Observer (FO) and ammunition at the Rearm Point (ammunition depot) and proceed to the coordinates XT5525.

I was the Aircraft Commander (AC) and in the Cu Chi traffic pattern while we were repositioning from the Bear Pit to the rearm point, we could see the yellow light of the flares over the 'contact' (a battle with the enemy). The flares were being dropped from the AC-47 (DC-3) Spooky Gunship just ten miles to the west of Cu Chi, between Trang Bang and the Saigon River.7

While we were rearming, I called DIVARTY (Division Artillery) and was given five separate artillery batteries from five different FSBs (Artillery Fire Support Bases) that were firing onto the same location that we were going to. Circling the infantry location with a grease pencil on my laminated tactical map. I then marked the artillery unit's locations and drew a black line from each directly to their target – our destination. The lines on the map were the artillery GTLs (Gun Target Lines). When I finished, it looked like a large wagon wheel with five spokes coming out of the hub, but no wheel rim.

AC-47 (DC-3) Spooky Gunship & Flare-ship;
"Puff the Magic Dragon"[4]

One of the artillery batteries firing in support of the Infantry that night was the artillery unit right behind our Little Bear living area on Cu Chi. From where we were sitting at the rearm point, it was clear that when we took off to the north and turned west towards the infantry location that we would be flying between two spokes of artillery Gun Target Lines (GTLs).

The co-pilot made a quick call to Clipper Control, the call sign for 25th Aviation Battalion Tactical Operations Center. Clipper Control would act as our flight following for the night.

After the aircraft was loaded with boxes of ammo and the Forward Observer climbed aboard, we took off and could clearly make out the "Bee Hive" in the flare-lit area over the contact. When there is a contact going on, aircraft of all sorts stack-up over the contact like so many illuminated honey bees circling a hive.

Generally, the lowest aircraft are Forward Air Controllers (FAC) who are adjusting artillery. There is usually also a FAC providing liaison between the USAF (United States Air Force), its DC-3 Spooky and F-4 Air-to-Ground bombers. Above them would be a couple of levels of Command & Control ships including the Battalion Commander, Regimental Commander and sometimes even the Division Commander. Above them at about six

[4] National Museum of the US Air Force (US Air Force Photo)

thousand to eight thousand feet (cloud cover permitting) was the DC-3 Spooky dropping flares and firing his miniguns around the Infantry perimeter. It was quite a surreal sight, bathed in the dim yellowish light of the parachute flares.

Then added to the mix was one or two pairs of UH-1C "Hogs" and alternately Cobra Gunship teams working the perimeter. On a clear night, looking out on the horizon, you could sometimes see up to three separate Bee Hives over different contacts in various parts of the AO (Area of Operation). Based on the coordinates you were given; just point your aircraft at your Bee Hive and go. It made it easy navigating at night if one of the contacts was where you were going, otherwise, if you weren't specifically going to that contact, you gave the BeeHives a wide margin.

The artillery gun target lines were a bit tricky to avoid on multiple contact nights. Some artillery units had multiple guns supporting more than one contact at the same time. Those particular artillery units were firing out of their Fire Support Base (FSB) in a couple of different directions simultaneously. That made safely flying around those FSBs even more challenging.

Occasionally, the USAF FAC would warn of approaching F-4s and provide their bombing run from north to south, or east to west, or on a specific magnetic track. The F-4s would dive into the target area aiming at the plumes of marker-rocket smoke that the FAC had fired and placed on target for them. After dropping their bombs or napalm, the F-4s would hit the after-burners and scream out from under the Beehive. You wanted to be well clear of the entry and exit flight path of the F-4s if you wanted to see another sunrise.

There was a lot happening on the ground with an "enemy contact," but there was also an awful lot going on in the air above the contact. You had to be very careful to avoid all other aircraft and keep track of where the fast-movers were at all times, not to mention Spooky firing his miniguns from above. We did that principally by getting the altitudes of all the other aircraft on station while arriving and advised the others of the altitude at which we would operate. It normally was a thousand feet separation vertically between aircraft, but I have seen five hundred-foot separations when overcast clouds forced the Beehive down to lower altitudes.

As we drew nearer, we could see artillery barrages blasting sections of the perimeter with mixed white and yellow explosions with red hot and white-hot shards of shrapnel flying out in all directions. There was a dust cloud that extended up a couple hundred feet above the ground that had been thrown up into the air from the repeated artillery explosions. The cloud would restrict our visibility but it would also help restrict enemy's visibility when we were approaching and landing.

AC-47 Spooky hosing down an Infantry perimeter time-lapse photo[5]

There were some anti-aircraft tracers being fired towards the Beehive aircraft every few minutes, but not heavy. There was enough yellow light ten miles to the east of the contact that allowed us to easily see the ground clearly, better even than on a bright moonlit night. I chose to fly low level because we were going to land shortly and to stay below the artillery trajectories. The rice in the rice paddies was high that time of the year extending well above the dirt berms that enclosed each of the paddies. The yellow light on the rice fields looked golden, making a distinct contrast with the dark trees in the uncultivated areas.

When we selected the infantry frequency that we were given, we heard a Medevac aircraft off Cu Chi saying, "The artillery will have to be stopped on one side of the perimeter so we can get in." The infantry Radio-Telephone Operator (RTO) responded, "We can't stop the artillery! It's firing 'Final Protective Fires' (FPF). We have been in hand-to-hand combat twice tonight already on parts of our perimeter and they'll be all over us if we stop the artillery. You'll have to come in without stopping the 'arty." The Medevac pilot responded, "Regulations prevent us from going in during Final Protective Fires. Call us when you can lift the artillery on one side of the perimeter."

Final Protective Fire (FPF) is when the Infantry and enemy are in such close quarters to each other that the normal safety distance of impacting artillery rounds is waived. The rounds

[5] National Museum of the US Air Force (US Air Force Photo)

are brought onto the perimeter of the infantry unit. There's no room for error. The last and final stage of artillery is when a unit is being overrun by the enemy and then the artillery is called into their own location. You definitely don't want to be that LZ if that happens. Which is going to get you first, Charlie, or our own artillery? In the end, it doesn't matter. Dead is dead!

RTO - Outlined against an evening sky, a radio-telephone operator moves to his post in his unit's night defensive position. It is through the RTO that the commander keeps in contact with his unit during action. (Photo by SP4 Robert O'Hare)

Hearing the Medevac aircraft pilot's conversation with the Infantry RTO, I started slowing the aircraft speed from one hundred knots back to forty knots and turned off the navigation lights and rotating beacon so Charlie couldn't see us, at least I hoped he couldn't! Even though I was flying the aircraft, I keyed the radio and called the Infantry unit saying, "This is Little Bear 16, we are bringing in your ammo and we can take your wounded out with us when we leave." The RTO responded with, "Thanks Little Bear 16, let us know when you are making your approach and we'll put down some additional cover fire for you in that sector." "Roger" I acknowledged.

I then briefed the crew. "All of us have heard the conversations on the radio. There is probably a seventy percent chance that we will not make it out of the LZ alive tonight, but a hundred percent chance that the grunts won't make it through the night if we don't go in. So, here's what we're going to do; I will be flying the aircraft and the co-pilot will be handling the radios. We'll lock our shoulder harnesses now so we won't fall forward onto the controls if hit. Confirm?" "Shoulder harness locked," came the reply from the co-pilot. I replied with "Harness locked".

Continuing the briefing; "I'm trimming the aircraft into a forty-knot attitude now and I'll be flying against the trim pressure. If I get headshot or wounded and my hands come off the controls, the aircraft will pitch up into a forty-knot attitude and climb instead of diving into the ground. At any time, you see the aircraft pitch up to a forty-knot attitude, take the controls without looking over at me or asking if anything's wrong. If you hear me say "handing over" or anything unintelligible, it may be the last thing I say, take the controls without looking or asking. Is that Clear?" "Roger" said the co-pilot.

I went on, "We'll continue to fly slowly and gunner and crew chief; you have 'free fire' from this point on. The only friendlies are in the LZ, so if you see any activity, fire without asking permission or telling me about it. As soon as an artillery barrage goes off on this side of the perimeter, I will be pulling in all the power we have and going as fast as we can to get into the LZ before the next barrage goes off".

"Keep firing your M-60s until we're over the perimeter and you hear me say "Cease fire!" At which time you are to cease firing immediately no matter what is happening on your side of the aircraft; unbuckle your seatbelt. Move into the cargo compartment and start throwing the ammo boxes out while we're still in the air. I want as much of the ammo off the aircraft as possible before we touch down in the LZ. We have wounded to pick up so help the grunts get them onto the aircraft so we'll have minimum ground time. We'll fly out of the LZ in a different direction than we're going in. Are there any questions?" I heard "No sir" three times on the intercom, one for each of the other three crew members.

Within thirty seconds, a barrage of artillery went off right in front of us on the perimeter, and I pulled in all the power we had and began accelerating and said to myself, "please don't let there be a late round!" A late round occurs while firing artillery, for some reason a round does not fire out of a single cannon. Everything is quickly reset on the misfired cannon and the round fired. The late round lands on the target from thirty seconds to a minute after the main barrage. That would ruin our day for sure if we were over the perimeter and a late round came in on top of us. That was one of the risks that we were taking landing into a live contact. The threat comes not just from the enemy's anti-aircraft fire; it also comes from our own artillery.

Both the gunner and crew chief began firing their M-60s. I told the co-pilot on the intercom, "Tell them we're coming in from the Cu Chi side." The co-pilot called the Infantry and told them, causing the whole perimeter to light up with tracers from outgoing rounds with some of the rounds passing just below us. The airspeed quickly accelerated to ninety knots before I had to flare the aircraft, reducing the power rapidly to make the final approach into the LZ. As we passed the perimeter, I called out on the intercom, "Cease fire – cease fire!" The M-60s stopped firing immediately.

I could hear boxes shuffling and being thrown out of the aircraft as we flew over the perimeter in a nose-high deceleration attitude. The LZ had been a farmer's house and yard with hedges of bamboo and a building, but everything was bombed out to within a couple feet of ground level. The tracers of the Viet Cong (VC) were coming in all across the LZ and the outgoing tracers were streaking out in all directions. It looked like a maze of illuminated wire crisscrossing the ground with very little area that was not covered with tracers.

I was looking for a black spot where the tracers weren't crisscrossing, because there was something like a tree stump blocking the bullets in that area. I did not want to set down directly in line with tracers that we could see. That would be suicide. I spotted the triangular shaped point inside the LZ where it was dark without tracers crossing and put the aircraft down there.

As I touched down, the right skid rested on a sandbag bunker about two feet off the ground with the left skid suspended in the air over a fighting position (fox hole). I maintained enough power to balance the aircraft on the precarious perch as the last of the ammo boxes were thrown off the aircraft. The replacement FO dove off of the aircraft into the dirt like a competitive swimmer dove into water on a racing start. As the aircraft got lighter, I had to keep reducing the power to hold the aircraft down on the sandbags under the right skid.

I was feeling particularly exposed and vulnerable as I watched the incoming streams of tracers on either side of our aircraft along with tracers whizzing past us from the rear. I was wondering how long it would take for one of the enemy soldiers to spot us at any moment and start firing directly into the cockpit. There's no protection from frontal shots except our chicken plates (body armor); nothing over the face that would deflect a bullet. "I can't be worried about that." I said to myself, as I focused on keeping the aircraft as stable as possible balancing on the bunker.

Something caught my attention in the foxhole to the left, directly down below my door. It was a soldier firing his M-16 on automatic with his hands extended over his head. He didn't lift his head up to see where he was firing. He just fired into his area of responsibility blindly on automatic. When he emptied a clip of ammunition, he pulled the weapon down into the foxhole, ejected the empty clip out of his M-16, replaced it with a full clip of ammo and repeated the automatic firing over his head.

I realized that with bullets coming in from three hundred sixty degrees around him, he could get his head shot off from behind so he's keeping his head down to stay alive. I also realized that the FO who was originally in the LZ had more than likely died because he had no choice about keeping his head down, but had to raise his head out of his foxhole long enough to see and adjust the artillery in closer to the perimeter.

After about four clips on automatic firing without aiming, the grunt stopped as he was pulling his weapon down for reloading and looked up at me sitting about seven feet off ground. He had that incredulous look on his face like "what in the hell are you doing up there?"

I got it! Immediately, I felt even more conspicuous than before, realizing that only plexiglass separated me from the outside hostile world. All of a sudden, it felt like I was sitting in a lifeguard tower at the beach. Sure, I could see everything around me, but everything around me could see me too!

I recalled a conversation that I had had in an aircraft with a grunt on the way out to an insertion. He was sitting looking forward in the cargo compartment just behind the radio console when I turned and nodded at him. He then said, "I wouldn't do your jobs for love or money. I'm happy dug into the ground and I wouldn't ever want to be sitting above the ground like you do to re-supply us during a fire-fight." I told the grunt "I like my showers each day and not having to sleep in the mud, the blood and the piss, and I would not want to change places with you either." We each did what we were trained for and respected the others for what they did. That was the way it was.

I refocused on holding the aircraft steady, balancing on the sandbags under the right skid. I became aware that two wounded soldiers were being dragged over to the aircraft by grunts on the right side of the aircraft. The grunts were low crawling along the ground to stay below the incoming tracers. The crew chief finished throwing the last of the ammo boxes off, reached over some boxes on the ground outside the aircraft and grabbed the end of the stretcher lifting it and sliding it into the cargo bay. He then grabbed the second stretcher and pulled it into the aircraft.

I could feel and hear when bullets hit the aircraft. It sounded like a hammer hitting sheet metal. We were taking a couple hits in the aircraft but none so far through the windscreen. The crew chief said, "We're ready," which was repeated by the gunner. I told the co-pilot, "Start looking for a barrage to go off on the other side of the perimeter and when it does, tell them we're going out that way".

Within thirty seconds, an artillery barrage went off to our right rear and I pulled in all the power we had and turned the aircraft in that direction as the co-pilot told the RTO which way we were going. That side of the perimeter erupted with outgoing automatic weapons fire and hand grenade explosions. They were covering our departure and it felt good.

As we transitioned from a hover and hit translational lift (flying speed) over the perimeter a loud explosion with brilliant white light went off right under the nose of the aircraft, and my chin bubble was blown in with broken plexiglass hitting me the full length of my body. The clear visor on my helmet was down over my face and saved my eyes. The concussion of the explosion pushed the aircraft up abruptly like it had been punched in the gut. More explosions were going off around the aircraft as we continued to accelerate out of the LZ. I was instantly having difficulty seeing clearly having lost my night vision in the bright white flash of the blast. We narrowly missed a couple trees as I jerked the aircraft left and then back to the right staying low level.

I flew out about a half mile away from the LZ to get away from the artillery kill zone and turned right toward the east to work our way back down to Cu Chi. I remained low level to enable us to fly under the artillery gun target lines. I turned the aircraft over to the co-pilot, because of all the windblast that was hitting me in the face from my chin being bubble mostly blown out. I couldn't use the radio either, due to the wind noise so the co-pilot called Clipper Control, advised of the damage and asked them to have another aircraft ready for us in ten minutes after we came back from the hospital pad. We flew to the 12th Evac Hospital pad on Cu Chi first and delivered the wounded soldiers, then back to the Bear Pit. We parked the aircraft on the maintenance pad and climbed into another aircraft assigned to us by Little Bear Flight Operations.

The crew chief for the second aircraft wasn't on the flight line so I told the existing crew chief to come with us. Usually the crew chief flies with his assigned aircraft whenever his aircraft flies. After a quick pre-flight, we cranked and were airborne again heading over to the rearm point. The aircraft was again filled with wooden boxes of ammo and we again took off for the LZ. I flew the leg again this time, and decided to not go into the LZ the same way as before. We surprised Charlie the first time. If we did it the same way each time, then Charlie would have a surprise waiting for us.

After taking off, I stayed low level and flew left under the gun target line (GTL) of the Cu Chi artillery and then turned right towards the LZ on the other side of the GTL. Again, looking for an artillery barrage to go off on the perimeter I slowed the aircraft and reconfirmed with the crew that "It's the same briefing as last time." Three "Rogers" came over the intercom. Within a minute a barrage went off right in our sector and I pulled in all the power we had and increased speed right to the perimeter of the LZ while the co-pilot told the RTO we were coming in from the southwest. Our M-60s roared into action and over the perimeter, I called "Cease fire – cease fire!" causing the M-60s to go silent.

This time we landed in another spot defined by a lack of tracers on the opposite side of the LZ from before. This time I was able to get both skids firmly on the ground. We were facing the side of the perimeter that we had gone out earlier but I wasn't going out that way this time. There were no wounded soldiers to Medevac out from that sortie. About the time the ammo boxes were all unloaded by the crewmembers with some help from a couple of grunts on the ground, a barrage went off to our left rear.

When the crewmembers were back on the aircraft, I pulled pitch as the co-pilot told the grunts we were going out to the northwest. That side of the perimeter erupted in defensive automatic weapons fire as we flew out of the LZ taking a couple hits as we stayed low and accelerating over the enemy positions. You could hear the AK-47s and see the muzzle flashes in the dim yellow light as we zigzagged to miss trees and hopefully bullets.

Upon arriving back at the rearm point on Cu Chi, we were once again loaded with ammunition and set out for the LZ. The turnaround was getting faster as there were more ammo bearers to help with the boxes. There was also another artillery Lieutenant FO at the rearm point to ride out with us to the LZ. He told us after he climbed onboard that the first two FOs had been killed. "Crap!" I muttered. That meant that the guy we had taken into the LZ on the first sortie was already dead! I felt a deep sadness come over me. I switched off my hot mike and took a couple of deep breaths to keep from tearing up. It didn't work, so I pulled my clear visor down over my face.

The Lieutenant told us, "I'm the Brigade Commander's Aide from Headquarters. There were no other FOs available to send out and they have to have an FO tonight." He looked away from us with that thousand-mile stare. I knew that look and what it meant. He was wondering if this was his night to die. He was a trained FO as was I from Artillery OCS (Officer Candidate School) before I went to flight school. But going from pushing paper to pushing artillery on a few minutes notice is a huge jump!

"OK" I said, as I turned back to the front trying not to show any emotion. I flew the aircraft again on that sortie, not comfortable enough to turn the controls over to the co-pilot for the time being. That was the Aircraft Commander's choice to make and I don't recall any co-pilot's questioning it. It had nothing to do with the copilot. It had everything to do with keeping the situation completely under control on the razor's edge.

This time we went to the right and crossed under the GTL of the artillery firing from the Saigon River near the Phu Cong Bridge, positioning us for a final approach from north to south into the LZ. On the way into the LZ this time we took some anti-aircraft fire but no hits. Once again, we unloaded rapidly and loaded one wounded soldier along with the two FO's bodies, both with head wounds.

We had to wait for an artillery barrage in order to fly out a different direction than we had before. When the barrage came, it was close to the front of the aircraft, just outside the perimeter. It startled all of us as white and red-hot shards of jagged steel ripped through the air just above the aircraft rotor blades. I immediately pulled in power and the copilot told the RTO that we were going straight ahead. We flew through the dust of the recent barrage.

As we were accelerating just out of the perimeter while flying between what was left of a couple of trees with most of their limbs and leaves stripped bare from shrapnel; either an RPG round or M-79 grenade launcher round hit a tree to the left of us. The flash of the explosion level with the aircraft illuminated the left side of the aircraft.

In the Little Bears, if your aircraft was damaged but you weren't wounded, you got back in the replacement aircraft and finished the mission. It was like the old days when you got thrown off of a horse; you got right back on the horse to keep you from getting scared. The policy seemed to work. It didn't keep me from getting scared of being killed, but it did keep me from getting scared of flying.

That time, we got a new crew chief with the new aircraft because Little Bear Flight Operations had anticipated and got the crew chief out to the flight line by the time we arrived. After a quick pre-flight, we cranked and were airborne again heading over to the rearm point while I briefed our new crew member in flight. Reloaded and back into the air, I gave that leg to the co-pilot to fly and I handled the radios. We low leveled around the left or south side of the LZ crossing under two gun-target lines this time before turning back right in towards the LZ. Again, timing the barrage, we flew in and dropped the ammo.

The RTO told us, "Little Bear 16, both Spooky and the Diamond Head gunships have gone off-station at the same time to rearm and refuel. After you drop the ammo, can you climb overhead and provide suppression around the perimeter as long as your M-60 ammo lasts?" "Roger, no problem," I replied. Everyone in the crew heard the radio transmission so I told the Co-pilot, "When you take off, we'll be light enough, just stay over the perimeter in a tight climbing spiral turn until we get to three thousand feet and we'll fire the guns." The co-pilot acknowledged and did just that.

The co-pilot adjusted the orbit about three hundred meters outside the perimeter and the crew chief began firing again around the outside of the LZ. When the grunts saw movement off their perimeter, they would direct our fire in those areas. When the crew chief's ammo can was empty on the right side, I took the controls and reversed the direction of the orbit to allow the gunner to fire his M-60. We finished off both two thousand round cans of M-60 ammo before flying back to Cu Chi for another rearm. We didn't refuel except every third or fourth sortie that night because of the short distances involved and to save the time between emergency resupply sorties. The lighter weight also made the aircraft more responsive and maneuverable.

We flew twenty-eighty sorties that night; fourteen sorties in and fourteen sorties out of that hot LZ. On the last sortie out of the LZ that night it was approaching five AM in the morning and still dark. I was flying the leg, low leveling directly toward the northern perimeter of Cu Chi over the rice fields when I heard a strange 'ssshhhhiiiisss' sound. I thought "What is that sound"?

My eyes abruptly opened to see that I had fallen asleep at the controls and the aircraft had settled into the rice while flying at a hundred knots low level. It was the rice hitting the nose of the aircraft that was making the strange 'ssshhhhiiiisss' sound. I immediately pitched the aircraft up and pulled in power to keep us from hitting one of the dirt berms that surrounded each paddy.

The middle of that devastation is where I landed 14 times and aged ten years in one night in Oct 1968. I took the picture the following week. The LZ we landed in was to the right of the two large bomb craters in the center.[6]

We flew thirteen hours that day with over nine hours of night. The co-pilot and I were written up for and were awarded the Distinguished Flying Cross (DFC) for the night's efforts.

That was one emergency re-supply mission. There were many.

2nd DFC: 28 Nov 1968, Dau Tieng, RVN - Tank Column Ambush: Trolling for Gators

One morning, I was tasked to fly to Dau Tieng. which was a thirty-minute flight to the northeast of Cu Chi. We were tasked to pick up an Armored Battalion Commander and proceed as he directed. After picking the Lt Colonel (LTC) at Dau Tieng, we flew to the west to locate one of his tank companies that had been ambushed while moving along the road through an old Michelin rubber plantation. When we arrived on site, I saw that the rubber

[6] Webb family photo Vietnam 1968

trees had been Rome Plowed (bull-dozed) back about fifty meters off either side of the road, but most of the tree debris was lying approximately knee high over the open ground. The tanks had stopped on the dirt road adjacent to the rubber trees and had turned perpendicular to the roadway facing the trees to their north. The tank column covered the entire length of the rubber tree stand.

Tracers were flying in all directions. The gunners atop the tanks were firing into the rubber trees to their north with their .50 cal machine guns, or right side of the road in their direction of travel. Pale green tracers (AK-47 rounds) were bouncing off of the tanks, being fired from within the plantation. After we arrived, the shooting stopped on both sides for a few minutes, during which time the Battalion Commander spoke with his ground tank commander on the radio.

The tank column had been dispatched from Dau Tieng to provide badly needed armor support for an Infantry unit currently in contact with a larger North Vietnamese Army (NVA) force further to the west. The armor needed to arrive at the Infantry's location before nightfall to fulfill their objective. They could not afford to allow themselves to be distracted with shooting it out with every last Viet Cong or NVA soldier that had ambushed them along the route.

In the briefing to us, the Lt. Colonel stated that he was concerned that the ambush could have been a coordinated, deliberately planned tactic by the NVA to delay and keep the armor column from arriving at the primary main contact in time, thereby giving the NVA the advantage in the main attack to the west. In line with that assessment, the Commander decided to move the column of tanks along the road and see if the enemy resumed firing.

The lead tank turned ninety degrees left in line with the dirt road and started to move forward when it was hit with a glancing RPG round (Rocket Propelled Grenade), but caused only superficial damage to the tank. That resulted in all the tanks firing their cannons and .50 cal machine guns into the rubber trees while the lead tank turned back to his right to face the rubber and resume firing. After about five minutes the shooting stopped from the VC/NVA force who were dug into the ground between the rubber trees. The tanks ceased fire as well. After no activity for another ten minutes, the Lt. Colonel asked me if I would be willing to fly down low level in front of the trees, so he could spot any bunkers that the enemy was using.

I said "Sure, but make sure that your tanks don't blow us up if Charlie starts shooting at us!" He called the ground tank commander, explained his low flying plan and told the tank Commander to order his tanks to only to fire behind the chopper, in response to any hostile fire from the rubber plantation.

With that, I briefed the crew chief on the right side of the aircraft, to fire at will when we flew down low level in front of the rubber trees. I then briefed the gunner on the left side of the aircraft to stow his gun and hold fire on the first pass, as he would be facing the tanks. Both acknowledged and we dropped down to make the first pass from right to left in front of the tank column about fifteen meters from the tree line and about five to ten feet above the ground.

As soon as we passed in front of the first rubber trees, enemy AK47s opened up from dug in positions on the floor of the rubber tree plantation about fifteen meters back from the face of the trees. The Crew Chief opened up with his M-60 spraying rounds as we flew past. I had chosen to fly at a hundred and ten (110-120) knots, so most of the AK-47 rounds passed under the tail and behind the aircraft. Speed is Life!

The tanks opened up firing their cannons and .50 cal's into the rubber where the green tracers were coming from. After the first run, I pulled steeply up climbing three hundred feet before completing a tight turn and dove steeply regaining speed as we dove back down to low level in front of the tree line a second time. I quickly reminded the Crew Chief, now on the side of the tanks, to stow his gun and to hold fire; and the gunner, now on the rubber tree side, to fire at will once in front of the rubber trees. The Lt. Colonel called his tank commander for a cease fire, and reminded them to only fire behind the chopper.

Again, we drew fire from the rubber trees and again the tanks fired their cannons into the rubber behind us, on the locations where the green tracers were coming from. On the third pass, there were no AK-47 tracers out of the rubber trees. Again, on the fourth and final pass, we didn't draw any fire. The Lt. Colonel asked me to climb to altitude so he could watch his tanks continue their road march. We climbed to fifteen hundred feet and all the tanks turned on the roadway facing west and started moving out with no further contact from the enemy.

Flying back to Dau Tieng, the Lt. Colonel thanked us for helping to get the Tank Company moving, so they could make their intended rendezvous to the west before dark. We dropped him off and proceeded to Cu Chi for the day. We were awarded a DFC for our troubles, which was a bit of a surprise, as I didn't expect anything.

Like the saying goes; *"God looks after aviators, kids and fools."* And for us, on that day, we had triple indemnity!

"Yea, though I walk through the valley of the shadow of death, I will fear no evil: for thou art with me; thy rod and thy staff they comfort me." (Psalm 23:4)

3rd DFC: 20 March 1969, Cu Chi, RVN - Last Emergency Resupply Mission

Two months before my year in Vietnam was completed in late May 1969, I was the aircraft commander of a Counter Mortar Standby crew on 20 March 1969. We were awakened by the field phone ringing around one in the morning. I was told that one of our crews had been shot up on an emergency resupply mission to the north of Cu Chi. Our mission was to deliver the ammunition that the previous crew were not able to deliver to the Infantry unit while in enemy contact. The Aircraft Commander of the stricken flight was W.O. Gerald Beaudoin, who had been wounded in the eye by antiaircraft fire during the attempted emergency resupply. Consequently, the approach had been aborted and the ground unit was still seriously low on ammunition.

Within a few minutes, we were cranking the aircraft and heading over to the rearm point to pick up the ammunition and other supplies. A Lt. Colonel, the Battalion Commander for the unit that we were flying in support of, met us at the rearm point and told us that he was going to ride out with us to check on his units. Not all Battalion Commanders took that much personal risk and he certainly didn't have to go with us on an emergency resupply, especially where one aircraft had already been shot up and a pilot wounded.

We were loaded in about ten minutes and were on the way to the Infantry unit's location, which was only about five kilometers to the north-northwest of Cu Chi. Because of the artillery gun target lines, we had to leave Cu Chi to the southwest, then turn to the north and approach the field location from the southwest. We climbed to fifteen hundred feet and called the frequency and call sign given. There was no answer.

There was a high thick overcast that night that no star light or moonlight penetrated. Away from the lights of Cu Chi base camp, it was like my Dad used to say in his crude way with words; "It's darker than six inches up a coal miner's backside, a mile underground." That's dark! We couldn't see the ground nor a horizon, so I flew locked onto our flight instruments.

The co-pilot called the unit frequency once again and this time we received an immediate response from the RTO. The RTO told us; "Enemy activity has ceased for the time being, but be careful. The last aircraft that tried to come in here got shot up less than an hour ago."

The co-pilot acknowledged the transmission and requested that a strobe be put out. The RTO responded; "We don't have a strobe that works. All we have is one trip-flare left. We'll put the trip-flare in the middle of the LZ." The co-pilot acknowledged and within thirty seconds, an extremely bright white phosphorous trip-flare illuminated, looking like a small sun, it was so bright. There certainly was no doubting where the ground unit was located.

The bright light of the magnesium trip-flare was at our twelve o'clock position and I started a descent directly toward it. At about three hundred feet above the ground, I observed jagged tree tops without leaves jutting up through the bright light, which meant that we were descending to a point into the trees short of the LZ. I pulled in power, leveled out and re-established a steeper descent path clear of the trees into the LZ.

Passing fifty feet above the ground, I was getting blinded by the brilliant light of the flare and began wondering if the flare was going to burn out before we landed, or were we going to have to set down on top of it? Neither was a good option.

If the flare went out while we were still on approach, then my night vision would be so badly shot, that it would have been very difficult to make a safe landing in pitch darkness with no night vision left. Conversely, if the flare kept burning and we sat down astraddle it, then the intense heat could have set our fuel tank on fire or blown us up. The fuel tank in the Huey is located under the cargo compartment and covers the under-belly of the aircraft. Vents under the aircraft belly allow fuel vapors and leaking fuel to drip out underneath as well. Barbeque anyone? I decided that we would not set down on a burning flare, but off to the side, space permitting.

Instead of shooting the approach all the way to the ground which is what we usually did, I brought the aircraft to a five-foot hover, trying to figure out if there was any other place to land in the LZ rather than on top of the white-hot inferno. Fortunately for us, within a few seconds, the magnesium trip flare burned out and with our landing light on, I was able to put the aircraft on the ground quickly.

The crew chief and gunner began off-loading the ammunition, assisted by an infantryman on each side of the aircraft. In less than a minute, the aircraft was unloaded and we lifted off into the darkness. We were about five feet in the air, moving forward approaching the perimeter, when a bright flash and loud explosion went off under and behind the aircraft, right where we had been sitting in the LZ. The aircraft had not hit translational lift speed (22 kts) and the concussion felt like a strong punch under the belly of the aircraft, which propelled us slightly to the left in the direction of flight. I was not sure what damage we had taken, but getting the aircraft in the air and clear of the LZ was all that I was focusing on.

There were no caution lights and all the fluid levels were where they should have been, so we continued on and climbed to fifteen hundred feet. The Battalion Commander sitting in the back, told me over the intercom that he would like to check on another one of his units a short distance further to the north and provided the coordinates. I turned the aircraft to the north as the co-pilot looked at his tactical map and located the coordinates, which were about fifteen kilometers northwest of the LZ that we had just left.

We arrived overhead the second LZ as the Lt. Colonel spoke with his Commander on the ground via one of our FM radios. That unit had not been in contact with the enemy that night, so we were not requested to land at their location. We continued overflying the LZ to Cu Chi base camp and dropped Lt. Colonel at his Battalion helipad. We refueled and returned to the standby pad for the last few hours of the night. There were no other calls to launch that night.

Much of the personal recrimination that I have put myself through in the years since the war, has most likely been *survivor's guilt*, part of Post-Traumatic Stress Disorder (PTSD), which is a common condition among combat veterans. My wife thinks that it is merely the melancholy part of my personality, dredging up and rehashing past events. She may be right …. as usual. All I know is that I still carry the survivor guilt all these many years later.

Strangely, with as many lives that we saved and protected through the year of combat operations, I still feel guilty for having survived when others did not. As irrational as it sounds, *survivor's guilt* is something that almost all combat survivors wrestle with, whether they are willing to talk about it or not.

John flew over eighteen hundred combat flight hours in helicopters during two years in the Vietnam War. He attended night college courses and extension courses for most of his eight years with the Army, amassing three years of college credit in the process. In 1973, he left the Army at the rank of Captain and returned to civilian life. He battled nightmares, mild paranoia and survivor's guilt and explosive anger, all characteristics of Post-Traumatic Stress Disorder (PTSD). He went on to pursue his dream of becoming an airline pilot. John retired in 2013 with over 28,500 flight hours, with no accidents or incidents, during more than 46 years of flying military, private, and corporate aircraft, as well as commercial airliners.

John completed his Bachelor of Science degree from Florida International University in Miami, Florida in 1974 and a Boeing 727 Flight Engineer Certification on the GI Bill later that same year. He worked as a salesman and sales manager, first in Miami, Florida, then in Atlanta, Georgia, while flying light aircraft to keep his hand in aviation. John was finally hired by Braniff International Airlines as a Boeing 727 Flight Engineer in January 1979. John felt like he had 'died and gone to heaven', but after eighteen months, he was laid off, as the airline declined towards bankruptcy.

In 1980, while dealing with the loss of his airline pilot job and a difficult divorce, John turned his life over to Jesus Christ, becoming a committed Christian and most of the PTSD symptoms disappeared.

After Braniff's bankruptcy, John continued to fly commercially, persisting through two more airline mergers and three airline bankruptcies that he worked for, remarrying in the process. Finally, in early 1990, John was hired by an Australian airline that quickly merged with Qantas Airways. John flew as Captain on Boeing 737's for thirteen years throughout Australia and New Zealand. John later flew the Airbus A-330 wide-body as Captain internationally for nine years throughout the Asia-Pacific region, including Tokyo, Singapore, Hong Kong, Shanghai, Beijing, Mumbai, Bali, Auckland, Christchurch and Los Angeles.

During John's last year with Qantas, he flew the largest, heaviest, most complex and computerized passenger aircraft in the world at the time. He flew the Airbus A-380, with a capacity of 600 passengers, with twenty-four flight attendants and four pilots. The A-380 Super Heavy wide-body behemoth flew at a maximum takeoff weight of one million, two hundred fifty thousand pounds (1,250,000 pounds).

John finished his flying career of forty-six years in 2013, having flown with six different airlines in the United States and Australia. He logged just under 28,000 flight hours. John believes that his tenacity strengthened by his faith in God sustained him and his wife Rose Anne through his challenging aviation career. John continues flying small aircraft privately in Ocala, Florida where he and Rose Anne now live in retirement.

Captain John H. & Rose Anne Webb

26

Joe Kittinger
Colonel
United States Air Force

Korean War, Cold War, Vietnam War

During his noteworthy career, Colonel Joseph Kittinger was awarded two Silver Stars, five Distinguished Flying Crosses, three Bronze Stars with Valor device, two Purple Hearts, and 23 Air Medals! As Astronaut Nicole Mann aptly put it, "He has done just about everything in the skies and in space that any of us aspired to." Joe is a member of the Distinguished Flying Cross Society and he lives in Central Florida. He is not able to attend our meetings without a lot of difficulty so we see him only rarely. However, we want to salute him for his service and recommend to all aviation enthusiasts his book "Come Up and Get Me."

A native of Florida, after attending the University of Florida, Joe joined the newly formed US Air Force in 1949. Upon graduating from pilot training as it was called then, he was assigned to a fighter squadron in Germany flying the P-47, F-84, and the F-86 all in a three-year period. His next assignment was to the fighter test section at Holloman Air Force Base New Mexico. While there he became involved with several projects in preparation for the coming space age. In a group led by Dr. John Paul Stapp, Joe became the first man in space.

Joe Kittinger conducted the first human research of weightlessness in a T-33 jet trainer, followed by the F-94, F-89, F-100 and the F-104.

In June 1957 he made the first flight of 'Project Man High' flying a stratospheric balloon solo to 97,000 feet, a record that remained for a significant time. He was then assigned to the Aero Med Lab in Dayton, Ohio where he made three research stratospheric parachute jumps from 67,000, 67,500 feet and 102,800 feet a record that stood for 52 years. For the jump from 102,800 feet, he was presented the Harmon Trophy by Pres. Eisenhower. After that, following another stratospheric balloon flight to 86,000 feet with Bill White an astronomer, Kittinger was assigned to the Air Commandos where he flew combat in Vietnam in 1963-64 in B-26's. He followed that with a combat tour in 1966-67, in A- 26s, then took a final tour in 1971-72 where he was the squadron commander of the Triple Nickel squadron flying the Phantom F-4 aircraft. In March 1972, he shot down a MiG-21, in May 1972 he was shot down by a MiG-21, and ended up as a POW in the Hanoi Hilton until released in March 1973. He accumulated a total of 1000 combat flying hours with 483.5 combat missions.

Upon retiring from the Air Force in 1978, he stayed very active in aviation flying barnstorming in a 1929 New Standard aircraft conducting sky writing, banner towing and flying gas and hot air balloons all around the world.

He won the prestigious Gordon Bennett gas balloon race four times. In 1981 he made a flight in a Cessna 180 from Orlando, Florida to Salzburg, Rhodesia for a total flight time of 96 hours. In 1984 he made the first solo gas balloon flight across the Atlantic Ocean launching from Caribou, Maine and landing some 86 hours later in northern Italy all while setting several world records.

Joe was the capsule communicator in October 2012 on project Stratos where through Joe's coaching, Felix Baumgartner jumped from a stratospheric balloon from 128,000 feet breaking Joe's record set in1960.

Joe is a fellow in the Society of Experimental Test Pilots, A member of the National Aviation Hall of Fame, the Parachutist Hall of Fame, the Balloon Hall of Fame, the Florida Aviation Hall of Fame, and a medalist in the Explorers Club. in 2008 he was awarded the Smithsonian Air and Space Museum lifetime achievement award for his aeronautical achievements and contributions to aviation. in addition, he received the National Aeronautics Association Elder Statesman Award for his contributions of significant value to Aeronautics.

His military awards include the Silver Star with one Oak leaf cluster the Distinguished Flying Cross with six oakleaf clusters, the air medal with 24 oakleaf clusters, the bronze Star with two Oak leaf clusters, and the Purple Heart with one Oak leaf cluster. He has accumulated 16,800 hours of flight time in 93 different aircraft and 2000 hours in gas and hot air balloons. In addition to all of that, Joe has 101 parachute jumps including two emergency ejections from military jet aircraft.

Currently Joe is an aviation and aerospace consultant and is a speaker for the aviation speakers Bureau. He lives in Altamonte Springs Florida with his wife Sherry and he still flies aircraft and balloons. His book "Come Up and Get Me" describes his multifaceted aviation career.

Joe was our first man in space, set the record for a freefall jump from space at 103,800 feet- a record that stood for 52 years. Joe was a principal advisor to NASA and has received the Wright Brothers Award for 50 years of contributions in aviation.

A true pioneer in various aspects of aerospace activities, Joe is a living legend. The Distinguished Flying Cross Society, Villages Chapter would like to salute Colonel Joe Kittinger for his many contributions, for his example and for his leadership.

Hand Salute! Ready two.

AH-64 Apache

Armed and dangerous over the Tigris and Euphrates rivers, Iraq

The Boeing AH-64 Apache is an American twin-turboshaft attack helicopter with a tailwheel-type landing gear arrangement and a tandem cockpit for a crew of two. There are features for a nose-mounted sensor suite for target acquisition and night vision systems.

It is armed with a 30 mm (1.18 in) M230 chain gun carried under the aircraft's forward fuselage, between the main landing gear. There are four hardpoints mounted under the stub-wing pylons for carrying armament and stores, typically a mixture of AGM-114 Hellfire missiles and Hydra 70 rocket pods. The AH-64 has significant systems redundancy vastly improving survivability in combat.

One of the revolutionary features of the Apache was its helmet mounted display, the Integrated Helmet and Display Sighting System (IHADSS); among its capabilities, either the pilot or gunner can slave the helicopter's 30 mm automatic M230 Chain Gun to their helmet, making the gun track head movements to point where they look.

Source: Wikipedia
Photo: Courtesy of pilot Don Procter

SECTION 5: GLOBAL WAR ON TERROR

The War on Terror, also known as the Global War on Terrorism, is an international military campaign launched by the United States government after the September 11 attacks. The targets of the campaign are primarily Sunni Islamic fundamentalist armed groups located throughout the Muslim world, with the most prominent groups being Al-Qaeda, the Islamic State, the Taliban, Tehrik-i-Taliban Pakistan, and the various franchise groups of the former two organizations.

The naming of the campaign uses a metaphor of war to refer to a variety of actions that do not constitute a specific war as traditionally defined. U.S. president George W. Bush first used the term "war on terrorism" on 16 September 2001,and then "war on terror" a few days later in a formal speech to Congress. In the latter speech, George Bush stated, "Our enemy is a radical network of terrorists and every government that supports them." The term was originally used with a particular focus on countries associated with al-Qaeda. The term was immediately criticised by such people as Richard B. Myers, chairman of the Joint Chiefs of Staff, and more nuanced terms subsequently came to be used by the Bush administration to publicly define the international campaign led by the U.S. While it was never used as a formal designation of U.S. operations in internal government documentation, a Global War on Terrorism Service Medal was issued.

Source: Wikipedia – "War on Terror"
https://en.wikipedia.org/wiki/War_on_terror

27

Donald K. Procter
Chief Warrant Officer
United States Army

Operation Enduring Freedom

Donald K Procter

A native of Claymont Delaware, born on March 1964, I grew up accustomed to the four seasons, the easy lifestyle near Delaware Bay, and a constant parade of aircraft overhead from Philadelphia Airport. I attended Claymont High School and participated in extracurricular activities including music and contact sports. Following my graduation in 1983, I looked around for what to do next; aviation was never far from my mind.

As a kid growing up in the 1960s, the Vietnam War, the invasion of Panama, and eventually 9/11 were constant reminders that life was not easy. Freedom is not free! After High School, I wanted to serve my country, so I enlisted in the United States Navy as an "AD-J" Jet Mechanic on P-3 Orion's stationed at Naval Air Station (NAS) Willow Grove, Pennsylvania VP-64 & VP-66. After serving in the Naval Aviation Wing for four years, I knew that I wanted to become an Officer and Pilot.

Naval Air Station (NAS) Memphis TN (center Airman "AD-J" Procter)

With encouragement from friends and family, I found my way into the Delaware Army National Guard (DEARNG). I graduated from Officer Candidate School at Fort Benning, Georgia, in August 1988 as a 2nd Lieutenant. I then went on to Fort Rucker, Alabama, "Home of Army Aviation" where I graduated as a pilot in August 1990. After serving the DEARNG for 10 years, I applied to the New York Air National Guard (NYARNG) where I later became a Combat Search and Rescue (CSAR) pilot flying the HH-60G Helicopter attaining the rank of Captain.

106th Rescue Wing HH-60G Pave Hawk, Incirlik Turkey Air Base (Operation Northern Watch-Iraqi)

Fast forward four years I was flying a Night training mission along the south shore Long Island over the Atlantic Ocean. I hit the tanker (C-130) for inflight refuel recertification and then landed back at home base Francis S. Gabreski Airport in Westhampton Beach. This was just another uneventful flight on September 10th, 2001. The next morning, I woke up, turned on the TV, and realized that my life would never be the same. 9/11 was a turning point for the country that would lead to 20 years of conflict in the Middle East. I was glued to the television for hours watching the devastation in New York, Washington, D.C. and Pennsylvania as major icons of American vitality and stability crumbled and burned while over 3,000 Americans died. I was ready to answer the call. The next day, I was boots on the ground standing at the base of what used to be the North Tower in downtown Manhattan, New York City. I was one of the ground Officers in charge (OIC), responsible for security and patrolling of the downtown area during the crisis.

Standing on adjacent building to the Twin Towers.

On patrol in the devastated mall under World Trade Center #5

Just like every red blooded American, I wanted payback. It was at that moment that I knew what I had to do; I wanted to fly helicopter gunships. Soon after that, I resigned my Commission as an Officer in the Air Force and enlisted as a Warrant Officer (W01) in the United States Army. Upon graduation and the transition to AH-64D Apache Longbow Helicopter, I reported to the 101st Airborne Division where I deployed to Iraq in 2003 and again in 2005. It was 2010 when I once more deployed to a combat zone, this time in Afghanistan at the Shindand Air Base.

AH-64D Apache Longbow
Call sign: Card Game
Aircraft: Apache Longbow Attack Helicopter AH-64D.
Manufacturer: Boeing (VERTOL).

US Army Armament: 30 mm M230 Chain Gun; Hydra 70 mm and CRV7 70 mm air-to-ground rockets; AGM-114 Hellfire; crew of 2 (pilot, and co-pilot/gunner) Propulsion: 2 × T700-GE-701C turboshafts Max Speed: 227 mph Range: 300 miles Conducts rear, close, and shaping missions including deep precision strike. Conducts distributed operations, precision strikes against relocatable targets, and provides armed reconnaissance when required in day, night, obscured battlefield and adverse weather conditions. The AH-64 Apache is the Army's heavy division/corps attack helicopter. The AH-64D Longbow remanufacture effort incorporates a millimeter wave fire control radar, radar frequency interferometer, fire-and-forget radar-guided HELLFIRE missile and cockpit management and digitization enhancements. The combination of the FCR, RFI, and the advanced navigation and avionics suite of the aircraft provide increased situational awareness, lethality and survivability.

On the day I earned the Distinguished Flying cross, we were flying typical aerial gunship support to US and allied troops in a rugged canyon in Western Afghanistan. I was the AH-64D Copilot Gunner for Bravo Company, 4-4 Aviation Regiment, Task Force Comanche, 4th Combat Aviation Brigade in support of an Italian and Afghan Special Forces combat assault, and search and seizure operation on the 2nd of October, 2010 located in Kharaba-I-Khwaja Mu, near the rugged Hindu Kish mountains north of Kabul, Afghanistan, AKA-"Death Valley".

At first it was not clear what sort of hostile fire we might face. After some initial engagements, the enemy forces attacked friendly forces on the ground and air while the sun appeared over the horizon. We were engaged in 3½ hours of close air support, which included two suppressive fire engagements on enemy positions due to continued escalation of enemy fire on our ground forces. As Co-pilot Gunner, I managed the battle from the front seat cockpit talking to the ground force maneuvers commander while engaging enemy targets with the array of weapons systems on board. The bottom of the mountainous valley was at 100 feet while sides of the valley rose to 2,500 feet in a river canyon which was only ¾ to 1¼ miles wide. We were below the enemy at times and under hostile fire from both sides of the valley. It was difficult to determine where the fires were coming from, as enemy forces were dug in.

We found ourselves in a target rich environment. The flying was challenging in itself but managing to hit targets around, above, and below us in close proximity with friendly forces, while maintaining close coordination with the various support aircraft and keeping in touch with allied ground forces made for a busy day. All seemed to be going well until my wingman ran low on fuel at a critical time in the ground battle.

"For he shall give his angels charge over thee, to keep thee in all thy ways." (Psalms 91:11).

Two AH-64D Apache Longbows from Task Force Comanche, 4th Combat Aviation Brigade, 4th Infantry Division, flying side-by-side launch flares during a mission near Shindand Air Base Photo via U.S. Army.

The narrative in the Army Times read: "Procter, who was circling overhead with his wingman to provide security, moved to the canyon to cover the pinned troops. As he arrived, his wingman was forced to leave the area to refuel, leaving Procter alone to provide security for the men on the ground and the five aircraft arriving to airlift the troops from the canyon. Seeing the Afghan and Italian soldiers engaged with the enemy 450 feet away, Procter opened fire with his 2.75-inch rockets and 30 mm guns, putting an end to the enemy attack and allowing the rescue aircraft, a mix of Chinooks and Black Hawk helicopters to land."

"When in combat, it is not like the movies and/or training," wrote Procter in an email to Army Times. "Without thinking you do what you must to save and protect their lives. They're my brothers in arms."

During the battle, the enemy forces opened fire at the Apaches overhead along with the exfil Black hawk and Chinook helicopters. Enemy fires left the rescue helicopters bullet ridden, though still operational as they swooped across the valley amid enemy fire to evacuate the ground troops. The battle resulted in citations for five Distinguished Flying Crosses, awarded to the Air mission Commander (AMC), and the four apache pilots, along with 14 Air Medals for Valor and seven Army Commendation Medals, also for valor.

Narrative to Accompany the Award of the Distinguished Flying Cross To Chief Warrant Officer Three Donald K. Procter

Chief Warrant Officer Three Donald K. Procter, United States Army, distinguished himself through exceptional heroism in aerial flight while serving as an AH-64D Copilot Gunner for Bravo Company, 4-4 Aviation Regiment, Task Force Comanche, 4th Combat Aviation Brigade in support of an Italian and Afghan Special Force's combat assault and search and seizure operation on 02 October 2010 in Kharaba-I-Khwaja Mu, Afghanistan. After having provided over 3½ hours of close air support, which included two suppressive fire engagements on enemy positions due to continued escalation of enemy fire on the ground forces, CW3 Procter conducted a successful battle handover with a follow-on Attack Weapons Team (AWT) and broke station. Shortly thereafter, the ground force began to receive effective enemy fire from every direction at distances between 300 to 1000 meters and called for an emergency extraction from the objective, which was situated in the bottom of a mountainous 2,500-foot-deep river canyon which was only ¾ to 1¼ miles wide. After monitoring a call that one of the AH-64Ds on the objective had received small arms damage in providing suppressive fires, when directed to return to the objective area by the overall mission AMC, CW3 Procter immediately responded without hesitation. Upon arrival, despite his wingman having to depart single ship for refuel, CW3 Procter remained to provide security for the lift aircraft through the canyon to provide suppressive fires during the extraction. Upon observing all five lift-aircraft on the PZ taking effective enemy indirect and small arms fire from only 150 meters away, with disregard to his own safety, he immediately engaged confirmed enemy locations with 2.75" rocket and 30 mm gun engagements, silencing them and facilitating the successful emergency extraction of the ground forces. As a result of his actions, all 91 ground force soldiers were successfully extracted while only sustaining four wounded. CW3 Procter's professional ability and heroic actions were above and beyond the call of duty and ensured the survival of numerous friendly forces.

"A thousand shall fall at thy side, and ten thousand at thy right hand; but it shall not come nigh thee. Only with thine eyes shalt thou behold and see the reward of the wicked." (Psalms 91:7-8).

Donald Procter Deployed

Citation to Accompany the Award of the Distinguished Flying Cross To Chief Warrant Officer Three Donald K. Procter

For exceptional heroism involving aerial flight while serving as an AH-64D copilot gunner in support of an Italian and Afghan Special Force's combat assault and search and seizure operation on 02 October 2010 in Kharaba-I-Khwaja Mu, Afghanistan. with disregard to his own safety, CW3 Procter silenced multiple enemy forces WHILE PROVIDING SECURITY TO LIFT AIRCRAFT DURING AN EMERGENCY EXTRACTION. CW3 Procter's heroic actions are in keeping with the highest traditions of military service, and reflect great credit upon himself, his unit, and the United States Army.

SECTION 6: IN MEMORIAM

Several members of our Distinguished Flying Cross Society have passed on. The Distinguished Flying Cross Society, The Villages chapter wants to continue to recognize and remember those men for their contributions in their respective wars by posting these Tributes to their accomplishments and heroinism. For additional details of their contributions, please consult the first edition of *Faces of the Distinguished Flying Cross of Central Florida,* which can be obtained from any member of the Club.

Remember that at some time they put their life on the line and pitted their skills and abilities to help their fellow man. Respect the fact they have been tested to the limits of skill and ability to help someone else during combat.

National Distinguished Flying Cross Memorial, March Field Air Museum, Riverside, California
Designed by Major James Lund Champlin, USAF (Ret)

28

In Tribute

Thomas Paul (Tom) Rozga
Major
United States Marine Corps (Ret)

World War II, Korean War, Cold War

RIP Tom Rozga, B. December 1, 1922 – D. March 30, 2018

Tom Rozga served with distinction in WW II, Korea and the Cold War through 1966.

Tom Rozga flew the most daring and harrowing mission we are aware of. Qualified as a F4-U fighter pilot, he flew an OY-1 light observation plane in the Pacific theater. His first DFC was for a ***night*** mission over Iwo Jima in a tiny OY-1 where he flew directly above the explosions of Naval artillery and under the arc of allied mortar and field-artillery rounds, to try and identify Japanese artillery pieces as they were being brought up from caves and dug-in positions. The OY-1 had no instrumentation to give him his position or pinpoint locations on the ground, could barely see due to smoke on the battlefields. (GPS was still 70 years into the future). Against all odds during a "pretty bumpy ride" he was able to pinpoint Japanese artillery pieces, and call in accurate friendly fire on enemy guns. His heroic action and superb flying skills identified and attacked Japanese artillery and caused the Japanese not to fire their big guns for 5 nights, thereby enabling US forces to gain an upper hand on Iwo Jima.

Stinson OY-1: 185 hp Lycoming 6 Cyl. Engine
The OY-1 was wood frame with cloth-covered fuselage and wings, (aluminum in critical short supply was limited to engine cowling, tail cone and strut fairings), cloth and epoxies used were flammable, very thin, vulnerable to bird strikes, small arms and flak.
Stall speed at 38 knots helped ability to land on Iwo Jima.

Only six years after Tom Rozga's heroic flight over the night skies of Iwo Jima in one of the most amazing flights we are aware of (Chapter One), he found himself at war again, this time with the Marines in Korea. The contrast of flying around a tropical island could not be starker; Korea was bitter cold (–40 F). The field conditions were abysmal. Tom earned two more Distinguished Flying Crosses in Korea while flying similar missions, again in OY-1's, as a spotter in minus 40-degree weather. When he got time to sleep it was in a trench on the ground, 100 yards from the Chinese line. Tom's courage, determination, skill as a pilot and keen observation skills were able to bring accurate artillery and friendly fire on the Chinese forces to help turn the tide of battle.

Even 72 years later, the memories of that experience were too painful for him to recount. (In the interview with him, his head bowed, he was clenching and unclenching his hands, as he had done all those years ago to fight freezing hands, his eyes were filled with tears and Gloria quietly urged me to not ask him any more about Korea). God knows what nightmares he has lived with, how many lives he saved. We will never know the price he paid or the degree of his sacrifice for our freedom. The Tom Rozga we knew was a quiet-spoken, kindly grandfather with an impeccable record of integrity, honesty, and superb family values. He certainly set a high bar for courage in action.

Tom Rozga with "Lady Satan"

29

In Tribute

Kenneth Clifford Taylor
Lieutenant
United States Army Air Corps

World War II

RIP Ken Taylor, B. August 10, 1922 – D. December 21, 2018

Ken was born in Puyallup, Washington, August 10, 1923, grew up in Sequim, Washington. He joined the Army Air Force in 1943, completed flight training and volunteered for the Pacific. He flew 66 missions against Japan.

On 29 May 1945 Ken Taylor's B-29 was hit by ground fire on a daring daylight bombing raid on the key industrial city of Yokohama. As they began their bombing run, intense antiaircraft fire from the very heavily defended area scored hits on their bomber. Beneath the floor of the navigator's compartment; fire broke out, filling the aircraft with smoke. Despite this hazard they remained in formation, released their bombs on the target with devastating effect. After bombs away, with the cabin completely filled with smoke, they left the protective formation to avoid collision with another plane. The crew prepared to bail out, however after breaking through the floorboards and using all the fire extinguishers and all the drinking water they brought the fire under control. The electrical system was almost completely burned out, the aircraft was vibrating and shaking; they flew the crippled aircraft with exceptional skill 750 nautical miles over open Pacific Ocean to Iwo Jima and landed under the most difficult of conditions.

Boeing B-29 Superfortress,
long range bomber designed and built to strike the heart of Japan

An avid golfer, Ken had many friends across the country; he golfed with Arnold Palmer, "had to wait behind Bob Hope and Bing Crosby". He lived a long, happy life.

We celebrate his life, and salute his flying skills and courage in the Pacific during WW II.

30

In Tribute

†

Rudolph (Rudy) Martin
Captain
United States Army Air Corps

World War II, European Theater (1943 and 1944)

RIP, Rudolf Martin, B. July 1921 – D. December 2017

Rudy Martin was born in Rockcreek, Ohio, July 1921. As war broke out with Europe, he volunteered for flight training; he qualified on the B-17, and deployed to England in 1943 as a B-17 pilot. 2nd LT Rudolph (Rudy) Martin was a pilot flying B-17 bombers with the 384th Bomber group against targets in Norway, Switzerland and Germany in WW II during 1943 and 1944.

Rudy joined the famed 384th Bomber Group, flew 25 B-17 bombing missions in 1943 based at airfields in central and southern England. He earned a Distinguished Flying Cross for flying 25 missions under opposing enemy anti-aircraft fire and hostile fighter plane air-to-air gunfire. His aircraft was seriously damaged on at least six occasions, his squadron was turned back by strong enemy fighter opposition, he ditched in the English Channel when he had engine trouble with the number 2 and number 4 engines, could not climb to altitude or maintain adequate airspeed. Rudy did very serious flying over Europe. Readers can find more information at the 384th web site http://www.384thbombgroup.com/.

Boeing B-17 Flying Fortress

A measure of the man he was: Of all the possessions he acquired and worked diligently to preserve over his lifetime, through permanent changes of station and the inevitable downsizing as he got older, Rudi saved a photo of his crew in front of a B-17, their names, and each of their 25 Sortie Reports from his flights over Europe in WW II. The names of his men are in the mission sheets and carefully inscribed on the back of the photo. He was proud they all made it home alive; kept track of them for 20 years. He did not tout his own flying skills, but was proud he got them all home safely.

Rudy Martin and his B-17 Crew

We salute you Rudy for your skill as an aviator and for your leadership.

Fair Skies and 100 mile visibility!

31

In Tribute

Daniel (Dan) W. Shea
Captain
United States Army Air Corps

World War II

RIP Daniel Shea, B. December 14, 1924 – D. August 8, 2013

Dan was raised as "the only Irish in an all-German settlement" called Marathon City, West of Wausau, WI. Dan was born Dec 14, 1924, sang, played High School basketball and liked to use the library, in the Monastery, on the east side of town. Dan and his generation answered their Country's call quickly and without reservation after Pearl Harbor. Dan joined the US Army Air Corps and was assigned as a Navigator Bombardier in the 465 Bomber Group (BG) out of Pantanella Airfield, near Canosa, Italy, in April 1944.

We don't know exactly how many missions his crew of ten flew, but he was awarded the Distinguished Flying Cross for a raid into Germany. Dan remembered on several missions they were "protected by the *Red Tail* Fighter Squadrons," as he proudly stated. The Tuskegee airmen of the Red Tails had a reputation for never running from a fight and never leaving an escort unprotected. The Tuskegee Airmen overcame segregation and prejudice to become one of the most highly respected fighter groups of World War II. They proved conclusively that African Americans could fly and maintain sophisticated combat aircraft. The Tuskegee Airmen's achievements, together with the men and women who supported them, paved the way for full integration of the U.S. military. Dan remembered their courage under fire.

"Red Tail" Tuskegee Airmen escorting a flight of B-17 Bombers[7]

[7] Colin Parker created this painting depicting a "Red Tail" Mustang P-51C, piloted by a Tuskegee Airman, escorting a squadron of B-17f bombers flying a mission in Europe during WWII. Prints of this work are available for purchase via Fine Art America. Copyright Colin Parker, all rights reserved. Image provided here as fair use for educational purposes and to acquaint new viewers with Parker's work.

CPT Shea's crew was awarded the DFC for a difficult mission (like so many others) against a BMW tank factory; they dropped their bombs from 27,500 feet. Their flight was under heavy fire from anti-aircraft guns. Even before dropping the bomb load (successfully), the Liberators came under attack from enemy fighters, BF-109s and FW-190s. In order to drop their bombs accurately they had to maintain a precise heading, through intense enemy ground anti-aircraft fire and air-to-air gunfire Dan held steady on his bomb sites inflicting major damage on the BMW plant. By the end of the bomb run the B-24's controls were only partially responsive and its defensive guns were reduced, by extreme cold, to only the top turret guns and one firing nose gun. Dan and his crew persisted with sluggish controls, lacking proper defensive weapons, flying back over the Alps and limping back to Pantanella. Despite numerous hits, the stricken plane made it back to fly another day.

His story of courage, skill and honor is included in this book, by his fellow members of The Villages Chapter of the DFC Society as a tribute to all he did in war, how he lived his life in peacetime, and his leadership here on Earth.

We salute you Dan! Enjoy the blue skies and 100-mile visibility.

32

In Tribute

William (Bill) Brown

Lieutenant Colonel

United States Air Force (Ret)

Cold War, Vietnam War

RIP Bill Brown, B. June 1934 – D. March 2019

William Brown, a native of Pensacola FL, enlisted in the Army Air Corps in Jan 1946 and was selected for pilot training in June 1949. He was an aviation cadet until September 1950 at which point, he was commissioned as second lieutenant ready to begin his career as an Air Force pilot.

While assigned to Iceland and the 1971st A.A.C.S. squadron, in 1959 to 1960 CPT Brown was instrumental in getting a status of forces agreement changed for the benefit of the United States Air Force. When complications arose between Icelandic and US interests, CPT Brown was able to negotiate and secure from the Icelandic Civil Air Authority a promise to accept a proposed change. This was made in writing and CPT Brown personally carried the proposal to United States Air Force headquarters in Washington DC to obtain the necessary signatures releasing five officers and five non-commissioned officers back to the AACS for assignment elsewhere. (This actually did not take effect until 1962 but the change was made and approved in August 1960).

Captain Brown transferred to Hurlbert Field in January 1962 where he and his crew trained in the C-47 in preparation for transferring to Temporary Duty (TDY) to Bien How, Vietnam, Detachment 2. In September 1962 the detachment commander received a request for a night medical evacuation to the mountains of Da Lat, South Vietnam to extract a special forces NCO who had broken his leg. CPT Brown volunteered, he and his crew were successful in flying the injured NCO to Tan Sohn Nhut for proper medical treatment.

Captain Brown returned to Hurlbert Field in December 1962 for C-46 qualification training and to test a balloon pick-up system, later transferring to Langley AFB Virginia as a standardization/evaluation pilot for the Special Warfare Section. He returned to Vietnam as a C-130 gunship aircraft commander over the Ho Chi Minh trail interdicting truck convoys. Lieutenant Colonel Brown flew a total of 300 combat missions spread between Korea and Vietnam for a total of over 1100 combat flying hours.

Through his leadership and initiative, Bill Brown established the Distinguished Flying Cross Society, Villages Chapter in 2009, and helped it grow to 30 members. We would like to recognize his accomplishments as a young man, salute his skill as a C-130 pilot, and his induction into the National Aviation Hall of Fame. He retired to Bridgeton, NJ where he worked in the banking industry.

Thank you for your leadership. Blue skies and fair winds.

33

In Tribute

†

Francis W. Bloomcamp
Colonel
United States Air Force (Ret)

Vietnam War

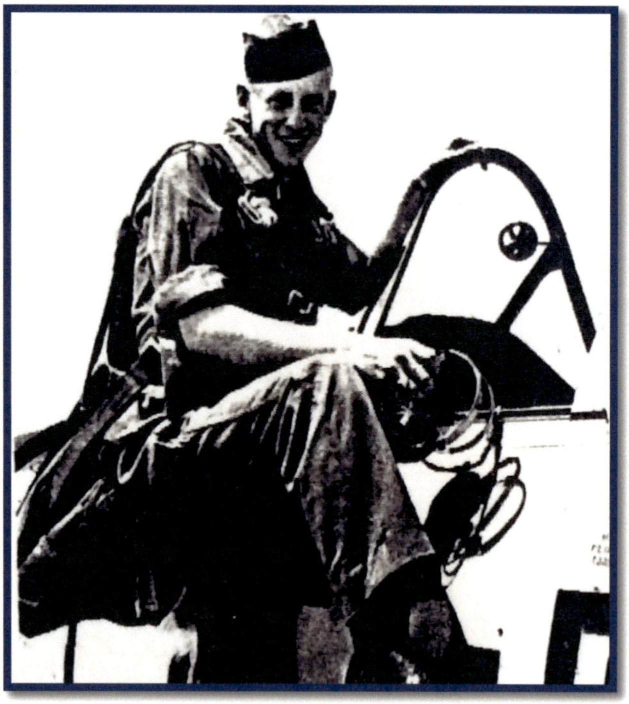

RIP Francis Bloomcamp B. July 5, 1932 – D. February 28, 2018

Colonel Francis W. Bloomcamp was born on 5 July 1932 in Brazilton, Kansas to Cletus and Esther Bloomcamp. After graduating from Helper High School in Brazilton, Frank joined the U.S. Air Force to pursue a career as a fighter pilot. He met many of the notable test pilots and fighter pilots of the day including Chuck Yeager and John Glenn. Air Force assignments during his 25-year career included tours in Europe, the Middle East, and numerous locations in the U.S. Frank served two tours in Vietnam earning a Silver Star for gallantry in action and two Distinguished Flying Crosses.

The citation for the first DFC states in part that Major Bloomcamp was the aircraft commander of a flight of F-4C's on 11 March 1967 against a major steel mill complex in hostile territory. Despite marginal weather and devastating ground fire and disregarding his personal safety, Major Bloomcamp pressed the attack and produced major damage to a vital part of the target complex. The second DFC was earned on 13 August 1967 when Major Bloomcamp was the aircraft commander leading a combat operational sortie against a vital railroad yard in North Vietnam. Despite hostile interceptor aircraft and under withering antiaircraft fire, Major Bloomcamp delivered his ordinance directly on target, resulting in the destruction of the railroad yard and numerous pieces of rolling stock. Both citations go on to state that the professional competence and aerial skill of Major Bloomcamp during these missions reflect great credit upon himself and the United States Air Force.

Returning to the U.S., Frank went on to Edwards AFB, California to lead the Air Force's F-15 fighter test program and later as the commander of the first F-15 fighter unit at Langley AFB, Virginia. He retired from the Air Force in 1978 as a full colonel when he joined the McDonnel Douglas Corporation, where for the next 10 years he served in international sales and marketing for the F-15. Frank enjoyed the outdoors and spent much of his retirement fishing and golfing at his homes in The Villages, Florida and Soldotna, Alaska. He was a generous and opinioned man who lived life on his own terms achieving greatness from humble beginnings. Colonel Francis W. Bloomcamp is buried at the Arlington National Cemetery.

Fair skies and 100 mile visibility; We salute you Frank.

Because We Flew

"Once the wings go on, they never come off whether they can be seen or not. It fuses to the soul through adversity, fear and adrenaline, and no one who has ever worn them with pride, integrity and guts can ever sleep through the 'call of the wild' that wafts through bedroom windows in the deep of the night.

When a good flyer leaves the 'job' and retires, many are jealous, some are pleased and yet others, who may have already retired, wonder. We wonder if he knows what he is leaving behind, because we already know. We know, for example, that after a lifetime of camaraderie that few experience, it will remain as a longing for those past times. We know in the world of flying, there is a fellowship which lasts long after the flight suits are hung up in the back of the closet. We know even if he throws them away, they will be on him with every step and breath that remains in his life. We also know how the very bearing of the man speaks of what he was and in his heart still is.

Naval Aviator

Because we flew, we envy no man on earth."

Army Aviator

~ Author Unknown

Air Force Pilot

EPILOGUE

The Distinguished Flying Cross Society
The Villages Chapter, Florida

Still active, making a difference, and having fun!

Back: Michael Gulick, Jack Deal, Dave Lehtonen, Wayne Lotsberg, John Webb, Dwight Moore.
Center: Norm Kasch, Richard Burns, Jack Hallett, Barry Spears, Dick Yocca, Gary Spooner.
Front: Barney Barnhart, Wayne Brown, Joe Finch, Don Procter, Dennis Derrig.

This book was a Club Project that brought us closer together, gave a better understanding of our lives, contributions and what each endured to be awarded a Distinguished Flying Cross. It was a fun project for everyone. New members were able to participate and add their stories to the original "*Faces...*" book published three years ago. Now we have a 2nd Edition, and with new members joining, we may have a 3rd offering in the near future.

As you reflect on the chapters you have read, go back and read between the lines. Try to understand the values of these young men as they willingly put themselves in harm's way to protect and defend their fellow men. And in so doing, defend our National interests, and protect the freedoms we all enjoy in this beautiful and capable nation established under God, with liberty and justice for all. Freedom often exacts a severe price. As long as young men and women are willing to pay that price we will continue to prosper.

We highly recommend this documentation and interview process to other Organizations and communities, to enable recording of their members' histories and contributions for future generations to learn from us and our core values.

The Distinguished Flying Cross Society, The Villages Chapter, Florida has regular meetings open to the public. If you are in the area try to attend one and meet some of the heroes featured in this book.

For additional information on acquiring a copy of this book, or attending a book signing event, reach out to any member of this Chapter. The hard-cover book is available from any Chapter member. The soft-cover and Kindle edition can be ordered from Amazon.com.

"Greater love hath no man than this, that a man lay down his life for his friends." (John 15:13)

TAPS: THE FINAL FLIGHT

Three slow drum beats signaled the extinguishing of lights going back to before the American Civil War that soldiers simply referred to as "Taps".

The Bugle Call for "Taps" was arranged in its present form by Medal of Honor recipient Union Army General Daniel Butterfield in 1862. Butterfield's bugler was the first to sound the new call and within months, "Taps" was adopted by both the Union and Confederate forces.

"Taps" was officially adopted by the United States Army in 1874 and went on to become an integral part of every military funeral since 1891.

Since "Taps" is a Bugle Call and not a song, no lyrics were originally associated with it. However, over the years, the following has been penned by authors unknown as fitting lyrics to this poignant and beautiful call.

Day is done, gone the sun
From the lakes, from the hills, from the sky
All is well, safely rest
God is nigh.

Fading light dims the sight
And a star gems the sky, gleaming bright
From afar, drawing near
Falls the night.

Thanks and praise for our days
Neath the sun, neath the stars, neath the sky
As we go, this we know
God is nigh

Dedicated to all the unsung heroes of former wars who performed their missions admirably and spoke quietly about the success of their units and importance of their crews.

Contact Us

Follow us on Facebook:

"The Distinguished Flying Cross Society, The Villages Chapter, Florida" (https://www.facebook.com/groups/636044717160301/)

The Villages Daily Sun Newspaper - Club News: under "Distinguished Flying Cross Club"

www.thevillagesdailysun.com
Address: 1100 Main Street, The Villages, FL 32159
Phone: 352-753-1119
Email: dailysun@thevillagesmedia.com

The Distinguished Flying Cross Society (www.DFCSociety.Org)

https://www.dfcsociety.org
Select: About - DFC Society Chapters
(https://www.dfcsociety.org/pages/chapter-locations)
Call Us Toll-Free: 1-866-332-6332

Finch Publications Inc.

Address: 3528 Valiant Court; The Villlages FL 32163
Email: josephfinch3751@gmail.com

Made in the USA
Las Vegas, NV
03 November 2026

57425086R00195